POWER TRAINS

A service, testing, and maintenance guide for power trains in off-road vehicles, trucks, and buses

FUNDAMENTALS OF SERVICE

PUBLISHER
DEERE & COMPANY
JOHN DEERE PUBLISHING
One John Deere Place
Moline, IL 61265
www.JohnDeere.com/publications

Fundamentals of Service (FOS) is a series of manuals created by Deere & Company. Each book in the series is conceived, researched, outlined, edited, and published by Deere & Company, John Deere Publishing. Authors are selected to provide a basic technical manuscript that could be edited and rewritten by staff editors.

HOW TO USE THE MANUAL: This manual can be used by anyone — experienced mechanics, shop trainees, vocational students, and lay readers.

Persons not familiar with the topics discussed in this book should begin with Chapter 1 and then study the chapters in sequence. The experienced person can find what is needed on the "Contents" page.

ACKNOWLEDGEMENTS:

Deere & Company also acknowledges help from the following: Allison Division, General Motors Corp.; American Gear Manufacturers Assn.; American Oil Co.; American Society of Agricultural Engineers; American Society of Mechanical Engineers; Bendix Corp.; Borg-Warner Corp.. Rockford Clutch Division; Cessna Products Corp.; Eaton, Yale and Towne Inc., Eaton Axle Div.; Gates Rubber Co.; General Motors Corp.; Goodyear Tire and Rubber Co.; Link-Belt/FMC; Lipe-Rollaway Corp.; National Fluid Power Assn.; Nuday Co.; Owatonna Tool Co.; Purolator Products Inc.; Rex Chainbelt, Inc.; Rockwell-Standard Co.; Rubber Manufacturers Assn.; Society of Automotive Engineers; Sundstrand Corp. Twin Disc Inc.; Vickers Inc., Division of Sperry Rand Corp.; Warner Electric Brake and Clutch Co.

Each guide was written by Deere & Company, John Deere Publishing staff in cooperation with the technical writers, illustrators, and editors at Almon, Inc. — a full-service technical publications company headquartered in Waukesha, Wisconsin (www.almoninc.com).

FOR MORE INFORMATION: This book is one of many books published on agricultural and related subjects. To download the latest catalog or for order information, please visit: www.JohnDeere.com/publications.

 We have a long-range interest in Vocational Education

Copyright © 1969, 1972, 1979, 1984, 1991, 2005, 2011, 2012. Printed in U.S.A. DEERE & COMPANY, Moline, IL/Eighth Edition/All rights reserved. ISBN 0-86691-377-7

This material is the property of Deere & Company, John Deere Publishing, all use and/or reproduction not specifically authorized by Deere & Company, John Deere Publishing is prohibited.

Power Trains

JOHN DEERE

FUNDAMENTALS OF SERVICE

Power Trains
ISBN-0-86691-377-7

FOS4008NC (2016)   (ENGLISH)

A service, testing, and maintenance guide for power trains in off-road vehicles, truck, and buses.

Deere & Company
PRINTED IN U.S.A.
11JAN16

Introduction

PUBLISHER

DEERE & COMPANY

JOHN DEERE PUBLISHING

One John Deere Place

Moline, IL 61265

http://www.JohnDeere.com/publications

FUNDAMENTALS OF SERVICE (FOS)

Fundamentals of Service (FOS) is a series of manuals created by Deere & Company. Each book in the series is conceived, researched, outlined, edited, and published by Deere & Company, John Deere Publishing. Authors are selected to provide a basic technical manuscript that could be edited and rewritten by staff editors.

HOW TO USE THE MANUAL: This FOS manual can be used by anyone — experienced mechanics, shop trainees, vocational students, and lay readers. The instructions are written in simple language so that they can be easily understood.

Persons not familiar with the topics discussed in this book should begin with chapter 1 and then study the chapters in sequence. The experienced person can find what is needed on the "Contents" page.

Each guide was written by Deere & Company, John Deere Publishing staff in cooperation with the technical writers, illustrators, and editors at Almon, Inc. — a full-service technical publications company headquartered in Waukesha, Wisconsin (www.almoninc.com).

POWER TRAINS

Power Trains is the definitive "how-to" book on power train systems of offroad vehicles, trucks, and buses – from showing you how to diagnose problems and text components to explaining how to repair the system.

This book discusses in detail:

- POWER TRAINS — HOW THEY WORK
- MANUAL TRANSMISSIONS
- HYDROSTATIC DRIVES
- INFINITELY VARIABLE TRANSMISSIONS (IVT)
- FINAL DRIVES
- SPECIAL DRIVES

And *MUCH* more.

ACKNOWLEDGMENTS

John Deere gratefully acknowledges the following people for their contributions to this manual:

Norm West, Agricultural Engineer, formerly with John Deere Waterloo Tractor Works.

Allison Division, General Motors Corp.; American Gear Manufacturers Assn.; American Oil Co.; American Society of Agricultural Engineers; American Society of Mechanical Engineers; Bendix Corp.; Borg-Warner Corp., Rockford Clutch Division; Cessna Products Corp.; Eaton, Yale and Towne Inc., Eaton Axle Div.; Gates Rubber Co.; General Motors Corp.; Goodyear Tire and Rubber Co.; Link-Belt/FMC; Lipe-Rollaway Corp.; National Fluid Power Assn.; Nuday Co.; Owatonna Tool Co.; Purolator Products Inc.; Rex Chainbelt, Inc.; Rockwell-Standard Co.; Rubber Manufacturers Assn.; Society of Automotive Engineers; Sundstrand Corp.; Twin Disc Inc.; Vickers Inc., Division of Sperry Rand Corp.; Warner Electric Brake and Clutch Co.

Copyright © 1969, 1972, 1979, 1984, 1991, 2005, and 2011. Litho in U.S.A. DEERE & COMPANY, Moline, IL/Eighth Edition/All rights reserved.

This material is the property of Deere & Company, John Deere Publishing. All use and/or reproduction not specifically authorized by Deere & Company, John Deere Publishing is prohibited.

ISBN 0-86691-377-7

FOS4008NC

PREFACE

Power Trains is the definitive "how-to" book on power train systems of offroad vehicles, trucks, and buses — from showing you how to diagnose problems and text components to explaining how to repair the system. And when we say "show you," we mean just that! Our book is filled with illustrations to clearly demonstrate what must be done… photographs, drawings, pictorial diagrams, troubleshooting charts, and diagnostic charts.

Instructions are written in simple language so that they can be easily understood. This book can be used by anyone, from a novice to an experienced mechanic.

By starting with the basics, the book builds your knowledge step by step. chapter 1 covers how power trains work. Chapters 2–11 go into detail about the different kinds of power trains and their working parts. Each chapter clearly discusses how to adjust and maintain its power train system as well as how to diagnose and test problem areas.

Contents

POWER TRAINS—HOW THEY WORK
INTRODUCTION	1-1
HOW A POWER TRAIN WORKS	1-2
HOW POWER IS TRANSMITTED	1-8
GEARS	1-11
BEARINGS	1-23
ADJUSTING THE GEAR TRAIN	1-27
POWER TRAIN SAFETY	1-30
SUMMARY	1-33
TEST YOURSELF	1-35

CLUTCHES
INTRODUCTION	2-1
TYPES OF CLUTCHES	2-2
DESIGN CONSIDERATIONS	2-4
CLUTCHES IN OTHER LOCATIONS	2-22
OTHER TYPES OF CLUTCHES	2-29
TROUBLESHOOTING	2-36
TEST YOURSELF	2-37

MANUAL TRANSMISSIONS
INTRODUCTION	3-1
TYPES OF SHIFTERS	3-2
SHIFT CONTROLS	3-18
CAM SHIFTERS	3-20
PARK LOCK	3-22
ADJUSTMENTS	3-22
GENERAL MAINTENANCE	3-22
TROUBLESHOOTING	3-23
TEST YOURSELF	3-24

POWER SHIFT TRANSMISSIONS
INTRODUCTION	4-1
ADDITIONAL BENEFITS	4-3
HYDRAULIC CLUTCHES	4-4
HYDRAULIC REVERSER	4-6
HYDRAULIC HI-LO	4-8
PLANETARY PARTIAL POWER SHIFT	4-11
FULL POWER SHIFT	4-14
PLANETARY POWER SHIFT	4-19
OPERATOR CONTROLS	4-19
COMPUTER CONTROLS	4-22
CONTROL VALVES	4-23
CALIBRATION	4-27
TROUBLESHOOTING	4-28
TEST YOURSELF	4-28

HYDROSTATIC DRIVES
INTRODUCTION	5-1
ADVANTAGES	5-1
HOW IT WORKS	5-2
BASICS	5-2
COMPLETE SYSTEM	5-7
OPERATION	5-8
HYDROSTATIC DRIVE AXLES	5-13
OTHER TYPES OF PUMPS AND MOTORS	5-14
MAINTENANCE OF HYDROSTATIC DRIVES	5-20
TESTING THE SYSTEM	5-20
SAFETY RULES	5-21
TROUBLESHOOTING OF HYDROSTATIC DRIVES	5-23
TEST YOURSELF	5-24

TORQUE CONVERTERS
INTRODUCTION	6-1
HOW IT WORKS	6-2
WHAT IS TORQUE CONVERTER STALL SPEED?	6-6
VARIATIONS ON TORQUE CONVERTERS	6-7
LOCK-UP TORQUE CONVERTER	6-7
TORQUE CONVERTER TRANSMISSIONS	6-7
TROUBLESHOOTING	6-10
TESTING THE TORQUE CONVERTER	6-13
TEST YOURSELF	6-13

INFINITELY VARIABLE TRANSMISSIONS (IVT)
INTRODUCTION	7-1
MECHANICAL OPERATION	7-4
HYDROSTATIC OPERATION	7-9
ELECTRICAL CONTROLS	7-13
COMPARISON WITH ANOTHER POPULAR IVT DESIGN	7-15
SUMMARY	7-18
TEST YOURSELF	7-18

DIFFERENTIALS
INTRODUCTION	8-1
RING GEAR AND PINION	8-1
DIFFERENTIAL LOCK	8-3
ADJUSTMENTS	8-9
DIFFERENTIAL STEERING	8-10

Continued on next page

Original Instructions. All information, illustrations and specifications in this manual are based on the latest information available at the time of publication. The right is reserved to make changes at any time without notice.

COPYRIGHT © 2012
DEERE & COMPANY
Moline, Illinois
All rights reserved.
A John Deere ILLUSTRUCTION ™ Manual
Previous Editions
Copyright © 1969, 1972, 1979, 1984, 1991, 2005, 2011

	Page
TROUBLESHOOTING	8-11
TEST YOURSELF	8-11

FINAL DRIVES
INTRODUCTION	9-1
STRAIGHT AXLES	9-1
PINION DRIVES	9-3
CHAIN DRIVES	9-4
PLANETARY DRIVES	9-5
DIFFERENTIAL STEERING	9-9
MECHANICAL FRONT WHEEL DRIVE (MFWD)	9-15
ADJUSTMENT	9-20
MAINTENANCE OF FINAL DRIVES	9-21
TEST YOURSELF	9-21

POWER TAKE-OFFS
INTRODUCTION	10-1
TYPES OF PTO	10-2
OPERATION	10-3
ASAE STANDARDS	10-4
DESIGN VARIATIONS	10-5
PTO BRAKE	10-11
CONTROLS	10-12
SAFETY RULES FOR PTOS	10-17
TROUBLESHOOTING PTOS	10-17
TEST YOURSELF	10-17

SPECIAL DRIVES
INTRODUCTION	11-1
BELT DRIVES	11-2
CHAIN DRIVES	11-7
RECIPROCATING DRIVES	11-11
CAM DRIVES	11-14
OVERLOAD PROTECTION DEVICES	11-15
SAFETY RULES	11-17
TROUBLESHOOTING OF BELT DRIVES	11-18
TROUBLESHOOTING OF CHAIN DRIVES	11-19
TROUBLESHOOTING OF GEAR DRIVES	11-20
TROUBLESHOOTING OF OVERLOAD MECHANISMS	11-21
TEST YOURSELF	11-21

DEFINITIONS OF TERMS AND SYMBOLS
DEFINITIONS	12-1

ANSWERS TO TEST YOURSELF QUESTIONS
ANSWERS TO CHAPTER QUESTIONS	13-1

POWER TRAINS—HOW THEY WORK

INTRODUCTION

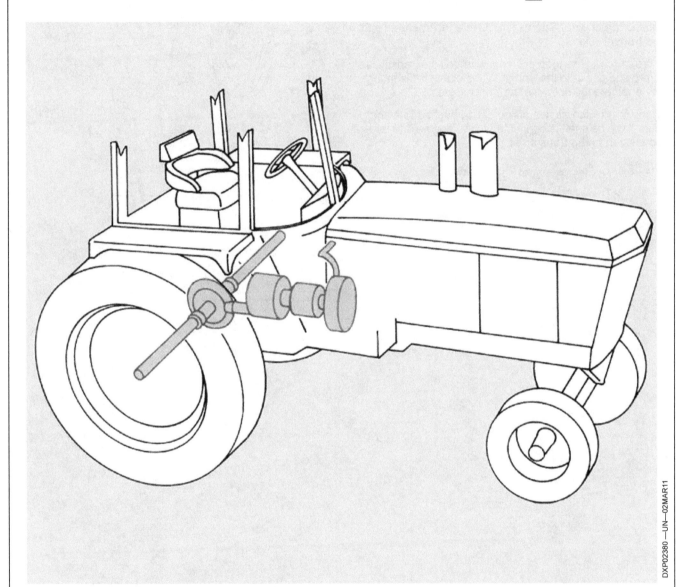

NOTE: *An animated display of how the components in a basic Power Train works is available on the Instructor Art CD.*

Engine power is transmitted to the drive wheels or output shaft of a machine by the power train. The power train does four jobs:

- Connects and disconnects power
- Selects speed ratios
- Provides a means of reversing
- Equalizes power to the drive wheels for turning

To do these jobs, five components are needed:

- Clutch — to connect and disconnect power
- Transmission — to select speeds and direction
- Differential — to equalize power for turning
- Final drives — to reduce speed and increase torque to the axle
- Drive wheels — to propel the machine

Continued on next page

POWER TRAINS—HOW THEY WORK

Additional components are also needed to help drive the machine, but these are the basic power train components.

Now let's follow the flow of power through a simple power train.

HOW A POWER TRAIN WORKS

The clutch connects and disconnects power. It is usually, but not necessarily, the first power train component behind the engine. The clutch may be incorporated into the transmission.

If the clutch is engaged, the engine drives the transmission to provide power to the wheels. If the clutch is disengaged, the engine does not drive the transmission.

Think of a clutch as two disks, each attached to a shaft (Fig. 1). One shaft is connected to the engine; the other is connected to the transmission.

A—Only One Disk Is Turning   B—Both Disks Are Turning

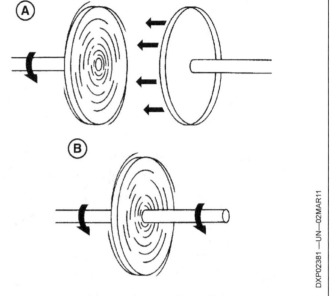

Fig. 1 — How the Clutch Works

Continued on next page

POWER TRAINS—HOW THEY WORK

Fig. 2 — Disk Clutch in Operation

A—From Engine
B—Flywheel
C—Clutch Disk
D—Pressure Plate
E—Spring
F—Cover
G—To Drive Wheels
H—Clutch Disengaged
I—Shaft from Engine Turning
J—Clutch Pedal Down
K—Drive to Wheels Not Turning
L—Clutch Disk and Plate Disengaged
M—Clutch Engaged
N—Clutch Pedal Up
O—Drive to Wheels Turning
P—Clutch Disk and Plate in Contact with Flywheel

If the two disks are not touching (disengaged), the engine turns one disk but not the other. No power is transmitted to the wheels (Fig. 2).

But if the two disks are pressed together (engaged), engine power is transmitted through the clutch to drive the machine. Both disks rotate as a single unit.

This is the principle of the disk clutch used on many machines. It is usually attached to the engine flywheel. Hydraulic pressure (not shown) or strong springs press the disks together. Pushing the clutch pedal down separates the two disks and disengages the clutch.

Certain machines have hydrodynamic or hydrostatic drives in which the clutch is eliminated. These drives will be covered later.

The transmission allows us to select an appropriate travel speed. Low speeds are needed for pulling heavy loads. The desired speed depends on conditions.

Also worth noting, the transmission enables us to use the appropriate engine speed. We might shift up and throttle back for better fuel economy under light loads, but drive the same speed at wide-open throttle in a lower gear for heavy loads.

The transmission is a system of gears.

Continued on next page

POWER TRAINS—HOW THEY WORK

Suppose we have a small gear with 12 teeth driving a larger gear with 24 teeth (Fig. 3).

When the first gear has made one complete revolution, it has gone around the equivalent of 12 teeth. The second one has gone around the same distance — 12 teeth — but this means only *one-half* a revolution for the larger gear.

As a result, the second gear and its shaft will turn at one-half the speed of the first gear and its shaft.

The point to remember is that when two gears are meshed, the smaller gear always turns at a faster rate.

This is the principle of the transmission — several combinations of gears are arranged so that we can select the speed we want.

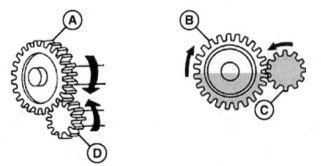

Fig. 3 — How the Transmission Selects a Speed

A—24-Tooth Driven Gear
B—Driven Gear Makes 1/2 Revolution
C—Driving Gear Makes 1 Revolution
D—12-Tooth Driving Gear

For low or first gear, a small gear on the input shaft drives a large gear on another shaft (Fig. 4).

This reduces the speed and increases the turning force, or torque.

Then a small gear on the second shaft drives a large gear on the output shaft that goes to the driving axle.

This reduces the speed and increases the torque still more, giving a higher gear ratio for low speed or heavy pulling.

For second gear, we can use the same first pair of gears as in low.

However, we disconnect the second pair of gears and drive through two other gears (Fig. 5).

These gears are arranged so that the larger one drives the smaller, so there is less overall speed reduction than in first gear.

For higher gears, the gear ratio is cut further by using other gear combinations. In fact, high gear is often an overdrive so the output is faster than the input.

A—Power from Engine
B—Power to Drive Wheels

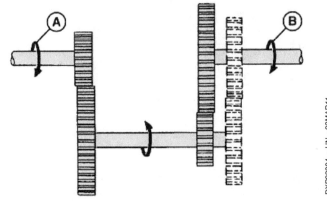

Fig. 4 — Transmission in Low or First Gear

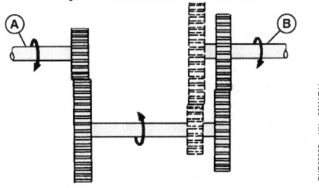

Fig. 5 — Transmission in Second Gear

Continued on next page

POWER TRAINS—HOW THEY WORK

Reverse gear is similar to first, except it incorporates an extra gear called a reverse idler (Fig. 6). The gear on the second shaft is not in mesh with the gear on the output shaft. Instead, both of them are in mesh with the reverse idler. This causes the output shaft to turn in the opposite direction from the input shaft.

All the gears are installed in a metal housing partially filled with oil to lubricate the gears and bearings.

The operator selects a gear by moving a shift lever in the driver's compartment.

In a power shift transmission, covered in chapter 4, mechanical shifting mechanisms are replaced by directing oil pressure to the appropriate elements.

Now let's complete the power train from the transmission to the drive wheels.

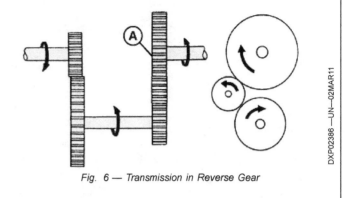

Fig. 6 — Transmission in Reverse Gear

A—Reverse Idler

The drive axle incorporates three sets of gears. First is the ring gear and pinion (Fig. 7). These bevel gears perform a 90° turn, connecting the drive shaft to the axle. They also provide a gear reduction, because a small gear is driving a large gear. Various ratios are available.

A—Pinion Gear
B—Ring Gear
C—Power "Turns the Corner"

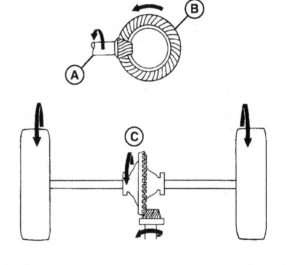

Fig. 7 — Ring Gear and Pinion for Drive Axle

Continued on next page

1-5

POWER TRAINS—HOW THEY WORK

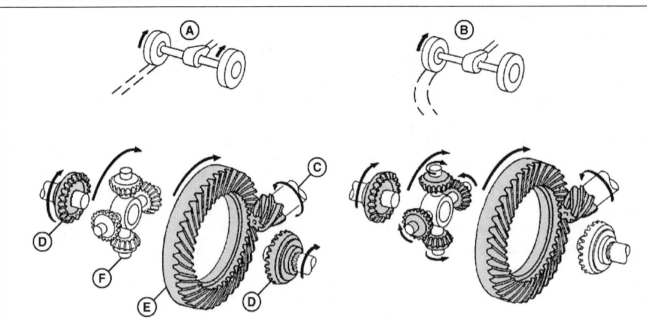

Fig. 8 — Differential

A—Driving in a Straight Line
B—Turning to the Right
C—Pinion Gear
D—Bevel Gear and Axle
E—Ring Gear
F—Bevel Pinions

Second is the differential, a set of bevel pinion gears (Fig. 8). The differential provides power to both wheels, even when one is turning faster than the other.

A differential would not be needed if the machine were always driven in a straight line. Both wheels would always turn at the same speed.

But the wheels must turn at different speeds when steering around a corner. Because the outside wheel travels farther, it must turn faster.

Bevel pinions are mounted in the ring gear. They are in mesh with side bevel pinions connected to the two axle shafts.

The left side of the diagram shows operation in a straight line. Both wheels are turning at the same speed. The bevel pinions rotate with the ring gear. Both side bevel pinions rotate at the same speed. It is as if the ring gear, bevel gears, and both side bevel gears were one solid piece.

The right side of Fig. 8 illustrates driving around a corner. The right wheel is turning more slowly. For explanation, let's say it is completely stopped.

The ring gear continues to turn as before. The bevel pinions are mounted inside the ring gear. However, the right side bevel pinion can't turn at all. This makes the bevel pinions rotate on their small shafts. Now the left side bevel pinion is driven at double speed, driven by rotation of the bevel pinions in addition to the normal operation.

However much one wheel slows down, the opposite wheel must speed up by the same amount. Power is delivered to both sides, however fast or slow they're turning.

If one wheel has poor traction and begins to spin, this could be a problem. A differential lock is often used to momentarily lock left and right sides together in slick spots.

Continued on next page

POWER TRAINS—HOW THEY WORK

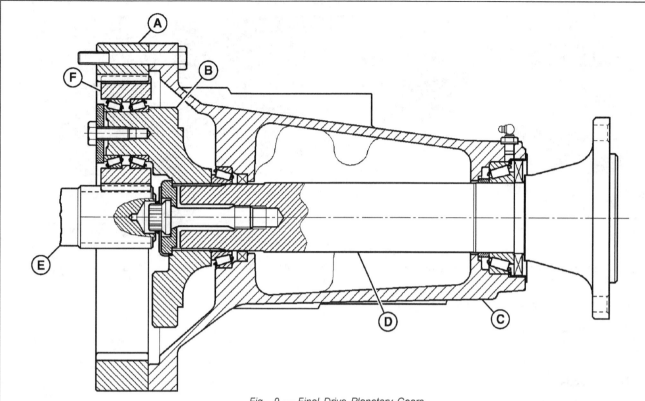

Fig. 9 — Final Drive Planetary Gears

A—Ring Gear
B—Planet Pinion Carrier
C—Axle Housing
D—Axle
E—Sun Gear (From Differential)
F—Planet Pinion (3 Used)

Third is the final drive (Fig. 9). Most farm and industrial machines use an additional set of gears on the axles to further reduce speed and increase torque. Bull gears and planetary gears are the main types. Because cars and trucks do not pull such heavy loads at such low speeds, they do not need final drives.

The axles, of course, rotate the wheels to make the machine move, which completes our basic power train.

Continued on next page

POWER TRAINS—HOW THEY WORK

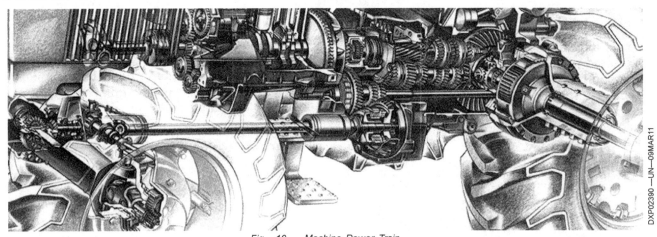

Fig. 10 — Machine Power Train

A complete power train in a modern machine is shown in Fig. 10.

For details on actual power trains, see these chapters:

- Chapter 2 — Clutches
- Chapters 3 and 7 — Transmissions
- Chapter 8 — Differentials
- Chapter 9 — Final Drives

HOW POWER IS TRANSMITTED

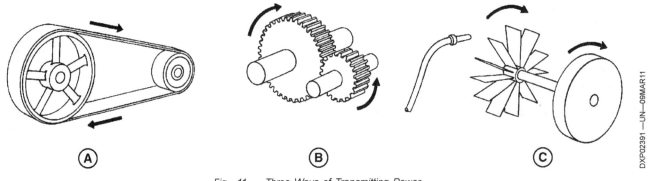

Fig. 11 — Three Ways of Transmitting Power

A—Friction B—Gears C—Fluids

We have seen how power is transmitted in a basic power train.

The clutch uses friction to transmit power, while the transmission uses gears. A third way to transmit power is by fluids, as in hydrostatic drives (Fig. 11).

So the three basic ways of transmitting power are:

- Friction (clutches and belts)
- Gears (in mesh)
- Fluids (water wheel principle or hydraulic motor)

Let's take a look at each type.

Continued on next page

POWER TRAINS—HOW THEY WORK

FRICTION DRIVES

Friction is caused by bringing together two surfaces made from materials that will transmit motion from one to the other. Slippage between the surfaces is a matter of engineering design. For our purposes, let's assume no slippage takes place.

Friction devices include clutches and belts.

BELT FRICTION

A belt friction drive uses two sheaves, or pulleys (Fig. 12). They can be separated by some distance if needed. A continuous belt provides the friction to transmit power from one pulley to the other.

Pulleys of different sizes can be used to increase or decrease speed.

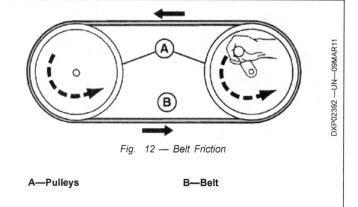

Fig. 12 — Belt Friction

A—Pulleys   B—Belt

Belt drives are covered in chapter 11.

GEAR DRIVES

Gears are the most common way to transmit power. When gears are in mesh, all slippage is eliminated. Friction and heat loss are low, so efficiency is high.

A chain drive (Fig. 13) is simply a variation of a gear drive. Rather than having gears in mesh with each other, we use sprockets connected by a chain. We can use sprockets of different sizes to increase or decrease speed.

Chains and sprockets are more positive than belts and pulleys. Chain drives are covered in detail in chapter 11.

A—Drive Sprocket   C—Driven Sprocket
B—Chain   D—Idler Sprocket (Adjustable)

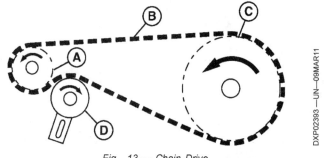

Fig. 13 — Chain Drive

Continued on next page

FLUID DRIVES

Fluid drives have been used for centuries in the form of water wheels and windmills. Modern examples include hydrostatic transmissions, torque converters, and propellers. The fluid is usually a liquid, but gas can also be used.

The old mill water wheel (Fig. 14) is turned by water falling into its buckets.

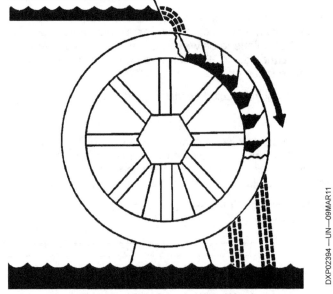

Fig. 14 — Fluid Drive In Water Wheel

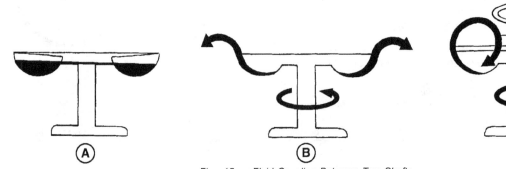

Fig. 15 — Fluid Coupling Between Two Shafts

A fluid clutch (Fig. 15) uses liquid to transmit power from one finned shell to another that is almost, but not quite, touching. Picture a doughnut or tire that has been sliced in two. Each half has radial fins that force the oil inside to spin along with the housing.

If you had only one half, as in A and B, centrifugal force would throw the oil out when the housing spins. Instead, the oil is thrown against the other half, where it strikes the radial fins and forces the output shaft to turn, as shown in C.

Torque converters, which use fluid clutches, are covered in detail in chapter 6.

The fluid clutch is not a "positive" drive. Speed of the output shaft depends upon load. You could stop the output shaft completely by holding the brakes, even with the input shaft turning at full speed.

Continued on next page

POWER TRAINS—HOW THEY WORK

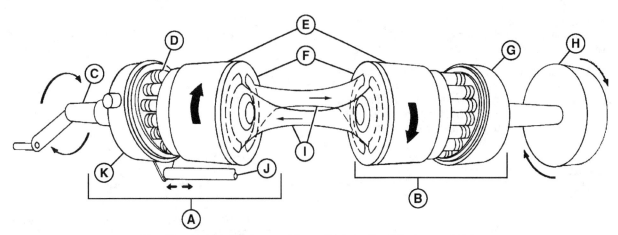

Fig. 16 — Variable-Displacement Pump Driving a Fixed-Displacement Motor

A—Pump (Variable Displacement)
B—Motor (Fixed Displacement)
C—Input Drive Shaft
D—Pistons
E—Cylinder Block
F—Valve Plates
G—Fixed Swashplate
H—Output Wheel
I—Oil Circuit
J—Swashplate Angle Control
K—Variable Swashplate

A hydraulic motor (Fig. 16) is a positive drive. By pumping oil into the mechanism, usually pistons that slide against a swashplate, we force something to move. If the output shaft can't turn, we could kill the engine.

Hydrostatic transmissions, which use hydraulic motors, are covered in detail in chapter 5.

GEARS

The basic elements of almost all conventional power trains are gears.

Gears are simply a means of applying twisting force — or torque — to rotating parts.

Torque is a function of force and leverage. Think in terms of a torque wrench (Fig. 17). The more force you apply, and the longer the lever, the more torque you are applying. In metric units, force is expressed in newtons, length in meters, and torque in newton-meters.

TORQUE	LEVER (D)	FORCE (E)
10 lb.-ft. (13.6 N·m) (A)	1 ft. (0.30 m)	10 lb. (44.5 N)
20 lb.-ft. (27.1 N·m) (B)	2 ft. (0.61 m)	10 lb. (44.5 N)
40 lb.-ft. (54.2 N·m) (C)	2 ft. (0.61 m)	20 lb. (89 N)

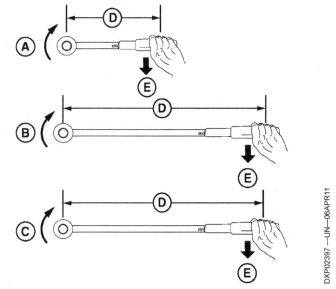

Fig. 17 — How Torque Varies with Force and Leverage

Continued on next page

POWER TRAINS—HOW THEY WORK

The principle is the same with gears (Fig. 18). The more force you apply against a gear's teeth, and the larger the diameter of the gear, the more torque you are applying.

So, if a large gear drives a small gear, the second one will have higher speed but lower torque than the first. A modern transmission provides a wide choice of speeds, so you can select the one that best fits the conditions. In simplest terms, all you are doing is choosing one combination of gear diameters.

A—Large Gear Driving Small Gear = More Speed but Less Torque

B—Small Gear Driving Large Gear = More Torque but Less Speed

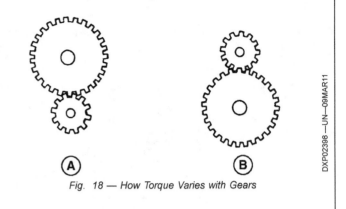

Fig. 18 — How Torque Varies with Gears

GEAR RATIOS

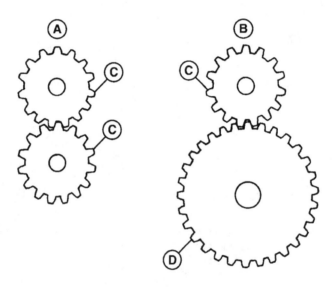

Fig. 19 — Gear Ratios

A—Ratio 1 to 1
B—Ratio 2 to 1
C—Speed 100 RPM
D—Speed 50 RPM

Gear ratio is a measure of the changes in speed and torque in the gear train.

The gear ratio between the gears in Fig. 19 is the ratio of the number of teeth on the lower gear to the number of teeth on the upper gear.

Continued on next page

TYPES OF GEARS

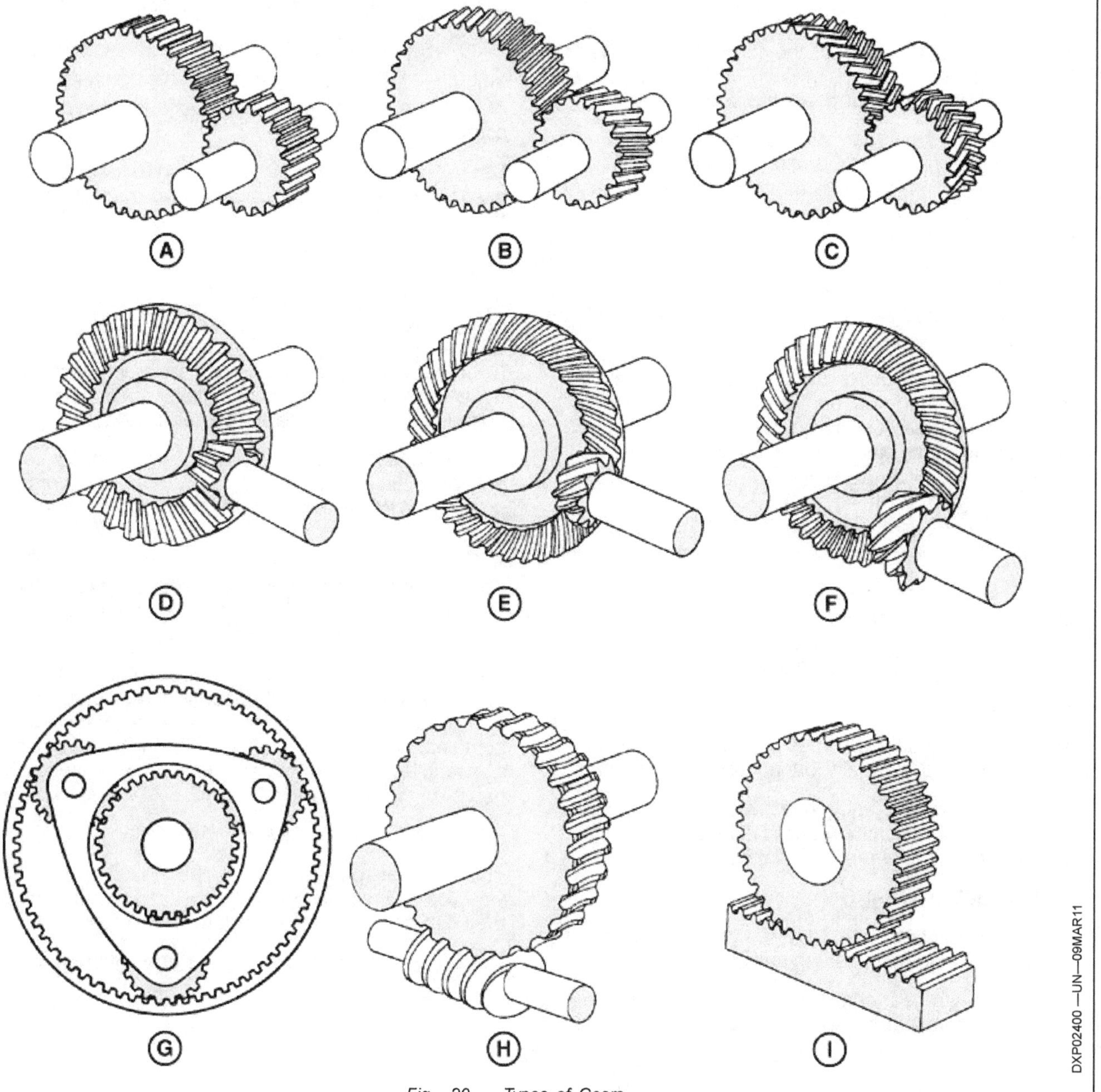

Fig. 20 — Types of Gears

A—Straight Spur
B—Helical Spur
C—Herringbone
D—Plain Bevel
E—Spiral Bevel
F—Hypoid
G—Planetary
H—Worm
I— Rack and Pinion

Various gear types are used to meet the demands of speed and torque.

Gears are normally used to transmit torque from one shaft to another. These shafts may operate in line, parallel, or at an angle to each other.

Meshing gears must also have teeth of the same size and design. And at least one pair of teeth must be engaged at all times. Some tooth designs allow for contact between more than one pair of teeth.

Gears are normally classified by:

- Type of teeth.
- Surface on which teeth are cut.

Some common types of gears are shown in Fig. 20.

Continued on next page

POWER TRAINS—HOW THEY WORK

STRAIGHT SPUR GEARS

Teeth are cut straight across the gear, parallel to the shaft. Normally, one or two pairs of teeth will be engaged at all times when straight spur gears are in mesh.

Uses: Straight spur gears are least expensive to machine. They don't produce end thrust on the shaft, so there's no need for bearings to carry a thrust load. They are used whenever there is no design requirement for features of more expensive gears.

HELICAL SPUR GEARS

Teeth are cut at an angle. This results in longer teeth, and more teeth being engaged at all times.

Uses: Helical spur gears are stronger and quieter. But they produce end thrust on the shaft, which must be carried by the bearings. Helical gears are used where strength, durability, and noise reduction are important.

HERRINGBONE GEARS

These have double helical teeth in a "V" pattern. There is usually a small gap in the center for oil movement.

Uses: Herringbone gears provide the advantages of helical teeth without the end thrust. Gears are self-centering. They provide quiet, high-speed operation on heavy loads such as large turbines.

PLAIN BEVEL GEARS

The plain bevel gear provides a change in shaft orientation. This is usually, but not always, a 90° turn.

Uses: The ring and pinion use bevel gears to drive an axle. Another common application is a right-angle gearbox in an implement's PTO driveline.

SPIRAL BEVEL GEARS

Like helical gears compared to straight spur gears, spiral bevel gears increase strength and reduce noise.

Uses: The ring gear and pinion are usually spiral bevel gears, because strength and noise are important issues.

An added benefit is that the angle of the spiral teeth can pull the pinion shaft, offsetting the push from the bevel angle and reducing load on the bearing.

HYPOID GEARS

This adds one more feature to spiral bevel gears. Instead of being on the centerline of the axle, the pinion shaft comes in lower.

Uses: Hypoid gears are mostly used to lower the drive shafts in automobiles. This eliminates the intrusive tunnel found in the floors of older cars.

PLANETARY GEARS

"Planet" pinion gears revolve around a "sun" gear in the center. The planet pinions also engage a ring gear on the outside. Planetary gears are strong, quiet, and versatile. Gear tooth load is spread among several meshes instead of just one. They exert much smaller loads on bearings, because outward forces are carried by the ring gear.

Uses: Planetary gears are used in many final drives and hi-lo or reverser assemblies. Many power shift or partial power shift transmissions also use planetary gears.

WORM GEARS

A worm gear is similar to screw threads, except the "threads" turn a gear. This provides a right-angle drive and a large gear reduction ratio.

Uses: Worm gears are usually used to convert high input speed to low output speed in a restricted space. Sometimes the output gear is only a sector instead of a complete circle, as in steering mechanisms. Certain early automobiles used worm gears instead of ring and pinion gears on the rear axle.

RACK AND PINION GEARS

This gear set converts straight-line motion into rotation or vice versa. Travel is limited by the length of the rack, so they can be used only for reciprocating motion.

Uses: Rack and pinion steering is one common example.

Continued on next page

POWER TRAINS—HOW THEY WORK

PLANETARY GEARS — HOW THEY WORK

Planetary gears are simple in design, but the variations in application can make them a challenge to understand (Fig. 21). The planet pinions (usually three) are installed in a planet pinion carrier. The planet pinions engage a sun gear in the center and a ring gear on the outside.

One component is restrained to prevent it from turning. This can be done by installing it rigidly in the housing or by engaging a brake to hold it for one speed selection.

Any of the three components can be restrained — ring gear, sun gear, or planet pinion carrier. Either of the other two can be the input, with the remaining component being the output. This provides six possible variations using one simple gear set.

A—Sun Gear
B—Planet Pinion Carrier
C—Ring Gear
D—Planet Pinions

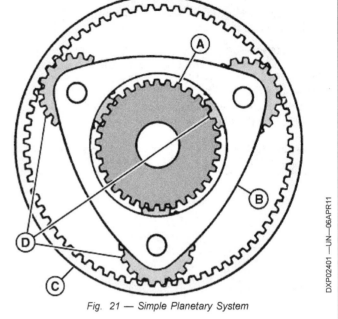

Fig. 21 — Simple Planetary System

SUN GEAR DRIVEN, RING GEAR RESTRAINED

Perhaps the simplest application is the final axle drive. The sun gear is the input, with power coming from the differential. The ring gear is installed rigidly in the axle housing. The planet pinion carrier is the output, splined to the axle shaft, which turns the wheel.

In this example, we rotate the sun gear (Fig. 22). This forces the planet pinions to "walk" around the stationary ring gear, rotating the planet pinion carrier. The carrier turns in the same direction as the sun gear, but much more slowly. The gear ratio depends on the relative diameters of the sun gear and the ring gear.

A—Power Output
B—Ring Gear Restrained
C—Sun Gear Driven

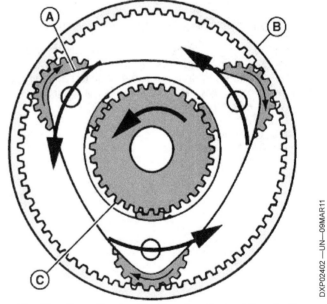

Fig. 22 — Power Flow in a Planetary Gear System When the Sun Gear Is Driven and the Ring Gear Is Restrained

Continued on next page

POWER TRAINS—HOW THEY WORK

SUN GEAR DRIVEN, PLANET PINION CARRIER RESTRAINED

We rotate the sun gear as before. This rotates the planet pinions, but they can't revolve around the sun because the carrier is restrained (Fig. 23). Instead, the rotating planet pinions force the ring gear to rotate. It turns in the opposite direction from the sun gear, but much more slowly. This provides a reverse gear.

A—Ring Gear Is Output (Reverse)
B—Sun Gear Driven
C—Planet Pinion Carrier Restrained

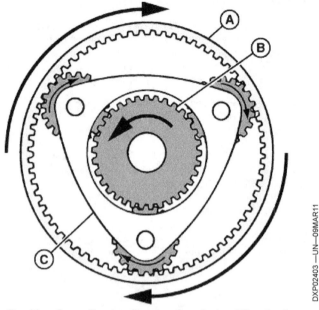

Fig. 23 — Power Flow in a Planetary Gear System When the Sun Gear Is Driven and the Pinion Carrier Is Restrained

RING GEAR DRIVEN, SUN GEAR RESTRAINED

Turning the ring gear forces the planet pinions to walk around the stationary sun gear (Fig. 24). This rotates the planet pinion carrier in the same direction as the ring gear, but more slowly. The reduction is not as great as in previous examples.

A—Ring Gear Is Input
B—Sun Gear Is Restrained
C—Planet Pinion Carrier Is Output

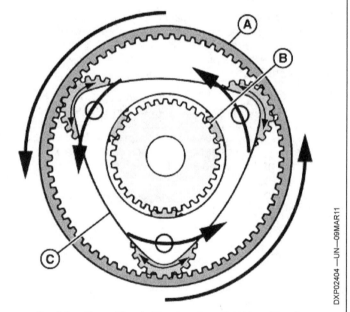

Fig. 24 — Power Flow in Planetary Gear Set When Ring Gear Is Driven and Sun Gear Is Restrained

Continued on next page

RING GEAR DRIVEN, PLANET PINION CARRIER RESTRAINED

Next, let's restrain the planet pinion carrier (Fig. 25). Now the sun gear is the output.

Turning the ring gear rotates the planet pinions, but they can't revolve around the sun because the carrier is restrained. Instead, the rotating planet pinions force the sun gear to rotate. It turns in the opposite direction from the ring gear, and much faster. This combination isn't used often because there is little need for high-speed reverse.

A—Ring Gear Is Input
B—Sun Gear Is Output
C—Planet Pinion Carrier Is Restrained

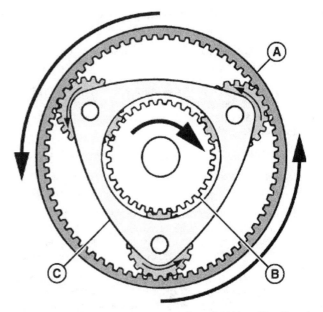

Fig. 25 — Power Flow in Planetary Gear Set When Ring Gear Is Driven and Planet Pinion Carrier Is Restrained

PLANET PINION CARRIER DRIVEN, SUN GEAR RESTRAINED

Fifth, suppose the planet pinion carrier is our input and the sun gear is restrained (Fig. 26). The ring gear is the output.

As the planet pinion carrier rotates, the planet pinion walks around the stationary sun gear. This forces the ring gear to rotate in the same direction as the input, but faster. This is very useful as an "overdrive" ratio.

A—Ring Gear Is Output
B—Sun Gear Is Restrained
C—Planet Pinion Carrier Is Intput

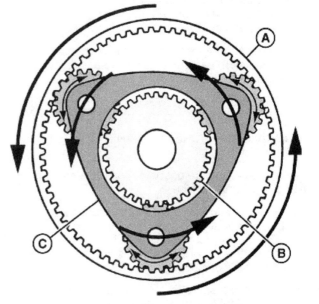

Fig. 26 — Power Flow in a Planetary Gear Set When Planet Pinion Carrier Is Driven and Sun Gear Is Restrained

Continued on next page

POWER TRAINS—HOW THEY WORK

PLANET PINION CARRIER DRIVEN, RING GEAR RESTRAINED

Finally, suppose the planet pinion carrier is our input and the ring gear is restrained (Fig. 27).

Now the planet pinions walk around the stationary ring gear. This turns the sun gear at a much faster speed. This provides another overdrive ratio, much higher than the previous example.

DIRECT DRIVE

There is yet one more possibility. Instead of restraining one component, we can lock two of them together.

If any two components can't move relative to each other, the whole assembly turns as one unit. It's as if we had a shaft instead of a gear set. Most transmissions use "straight through" direct drive as one of the gear selections.

In summary, it would be possible to produce seven speeds from the simplest planetary gear set — two reduced speeds, two overdrive speeds, two reverse speeds, and direct drive. Clutches and brakes can be engaged to turn or restrain the desired elements.

It would be surprising if anyone used all seven possibilities, but it's common to use a few. A simple planetary gear set provided the whole transmission for certain early automobiles. They used one reduction for low gear, direct drive for high, and another reduction for reverse.

With the addition of a torque converter on the input shaft, more modern cars have done the same thing for a two-speed automatic transmission.

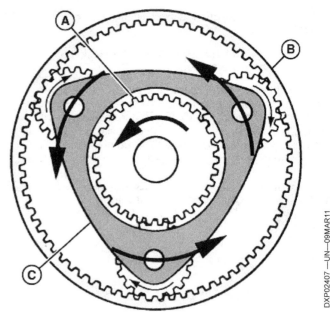

Fig. 27 — Power Flow in a Planetary Gear Set When Planet Pinion Carrier Is Driven and Ring Gear Is Restrained

A—Sun Gear Is Output
B—Ring Gear Is Restrained
C—Planet Pinion Carrier Is Input

ADDITIONAL VARIATIONS ON PLANETARY GEAR SETS

Even more variations are possible. Some are common in farm and industrial machines.

One thing we can do is to drive two components of the planetary gear set at the same time, but at different speeds (Fig. 28). No component is held stationary. The third component is the output, and it is driven at a speed determined by the other two.

In a power shift transmission, this provides an additional overdrive combination. In an infinitely variable transmission (IVT), it provides the infinite variability. In tracks steering, it controls relative speeds of the left and right axles.

A—Ring Gear Is Output
B—Sun Gear Is Driven at a Different Speed
C—Planet Pinion Carrier Is Driven at One Speed

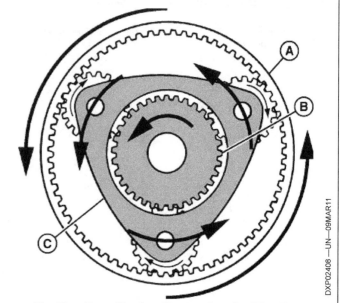

Fig. 28 — Power Flow in a Planetary Gear Set When Two Inputs Are Driven at Different Speeds

Continued on next page

POWER TRAINS—HOW THEY WORK

Sometimes it's helpful to use reverse idlers on the planet pinions (Fig. 29). This turns the output shaft in the opposite direction.

A—Planet Pinions
B—Sun Gear
C—Ring Gear

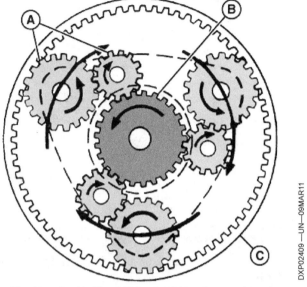

Fig. 29 — Double Planet Pinion Set Giving Reverse Speed

For even more combinations, we sometimes use "compound planetaries" (Fig. 30). Each planet pinion consists of two gears side by side. The two gears are different sizes, and they engage different size ring gears and sun gears. Power shift transmissions use compound planetaries.

A—B1 Sun Gear
B—B2 Sun Gear

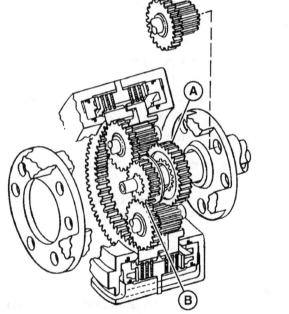

Fig. 30 — Speed Planetary Operation

Continued on next page

1-19

BACKLASH IN GEARS

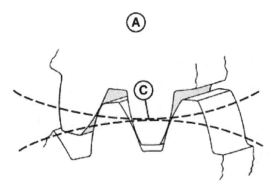 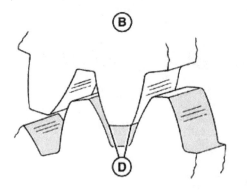

Fig. 31 — Backlash in Gears

A—Normal Gear Mesh
B—Too Much Backlash
C—Pitch Diameters
D—Backlash Occurs Here

Backlash is the clearance or "play" between two gears in mesh.

Too much backlash can be caused by worn gear teeth, an improper meshing of teeth, or bearings that do not support the gears properly.

Too much backlash can result in severe impact on the gear teeth. Broken gear teeth can result.

Too little backlash causes excessive overload wear on gear teeth. This could result in premature gear failure.

Fig. 31 shows normal gear mesh and gear mesh that permits too much backlash.

On the normal gears in Fig. 31, the clearance of the teeth at the pitch diameters is very small.

However, on the worn gears with too much backlash, forces cause a greater movement and a higher impact, which can break the teeth or at least cause the gears to bounce.

Backlash, endplay, and preload on gears and shafts is covered later in this chapter under "Adjusting the Gear Train."

GEAR WEAR

New gear teeth have slight imperfections, but they normally disappear during break-in as the teeth are oiled and polished. After that, the teeth should have a long service life.

However, when lack of lubrication or other factors cause a gear to fail, we can examine the failure and determine the cause.

The major types of gear tooth wear and failures are shown in Fig. 32.

NORMAL WEAR

This is the normal polishing of the gear teeth as they operate. The polished surface should extend the full length of the tooth from near the root (or bottom) to the tip of the tooth. Gears that are manufactured properly, well lubricated, and not overloaded or improperly installed will show this condition after many hours of service.

ABRASIVE WEAR

Surface injury caused by fine particles carried in the lubricant or embedded in the tooth surfaces. The causes are metal particles from gear teeth, abrasives left in the gear case, or sand and scale from castings.

SCRATCHING

This is often found on gears that handle heavy loads at slow speeds. This is caused by particles of metal flaking off the gears. Generally, it indicates the wrong gear design for the load. (Do not confuse this with scoring.)

OVERLOAD WEAR

If the contact surface is worn but smooth, the gears have been overloaded and metal has been removed by the sliding pressure. Continuous use will result in backlash and severe peening, which may be misleading as to the real cause of the wear.

ROLLING AND PEENING

Rolling is the result of overload and sliding, which leaves a burr on the tooth edge. Too little bearing support or too ductile a metal results in plastic flow of the metal due to sliding pressure. Peening is the result of backlash and the force causing a tooth to hammer on another with tremendous impact. In these cases, lubricants are forced out and metal bears directly on metal.

RIPPLING

This is a wavy surface or "fish scales" on the teeth at right angles to the direction of slide. It may be caused by surface yielding due to "slip stick" friction resulting from lack of lubrication, heavy loads, or vibrations.

Continued on next page

SCORING

This is caused by temperature rise and thinning or rupture of the lubricant film as from too heavy loads. Pressure and sliding action heat the gear and permit metal transfer from one tooth to the face of another. As the process continues, chunks of metal loosen and gouge the teeth in the direction of the sliding motion. The temperature rise here is slow and not as high as burning wear.

PITTING

Gear teeth should not show pitting. Very minute or micro-pitting can occur and would appear as a gray surface, which may advance slowly to an actual pitted condition. This condition is sometimes associated with thin oil film, possibly due to high oil temperatures.

SPALLING

Spalling is a common wear condition that starts with fine surface cracks and eventually results in large flakes or chips leaving the tooth face. Improperly case-hardened teeth are most often subject to this kind of damage due to the brittle nature of the metal. Spalling may occur on one or two teeth, but the chips may cause other damage to the remaining teeth.

CORROSION

Corrosive wear results in an erosion of the tooth surfaces by acid. The acid is formed by moisture combining with lubricant impurities and atmospheric contaminants. Generally the surfaces become pitted, causing an uneven surface and distribution of stresses, which lead to chipping and spalling.

BURNING

Burning is usually caused by the complete failure of lubricants or a lack of lubrication. During high stress and sliding motion, friction develops rapid heating and the temperature limits of the metal are exceeded. Burned gear teeth are extremely brittle and easily broken.

INTERFERENCE WEAR

This type of wear can be caused by misalignment of gears, which places heavy contact on small areas. Also, mating of two gears with teeth not designed to work together will cause interference wear. More than one wear pattern may show up, as at teeth tips and roots.

RIDGING

These are scratches appearing near one end of a tooth, especially on a hypoid pinion gear. This can be caused by excessive loads or lack of lubrication, or by the gear not properly heat treated during manufacture.

BREAKAGE

Broken teeth may be the result of many defects. Make a close study of the other teeth before judging the cause. Breakage can be caused by high impact forces or defective manufacture. To determine if breakage is due to overload or fatigue, examine the broken area closely. If the break shows fresh metal all over the break, an impact overload was the cause. If the break shows an area in the center of fresh metal with the edges dark and old-looking, the breakage was due to fatigue that started with a fine surface crack.

CRACKING

These failures tend to be caused by improper heat treating during manufacture. Improperly machined tooth root dimensions can also result in cracking. Most heat treat cracks are extremely fine and do not show up until a gear has been used for some time.

Continued on next page

POWER TRAINS—HOW THEY WORK

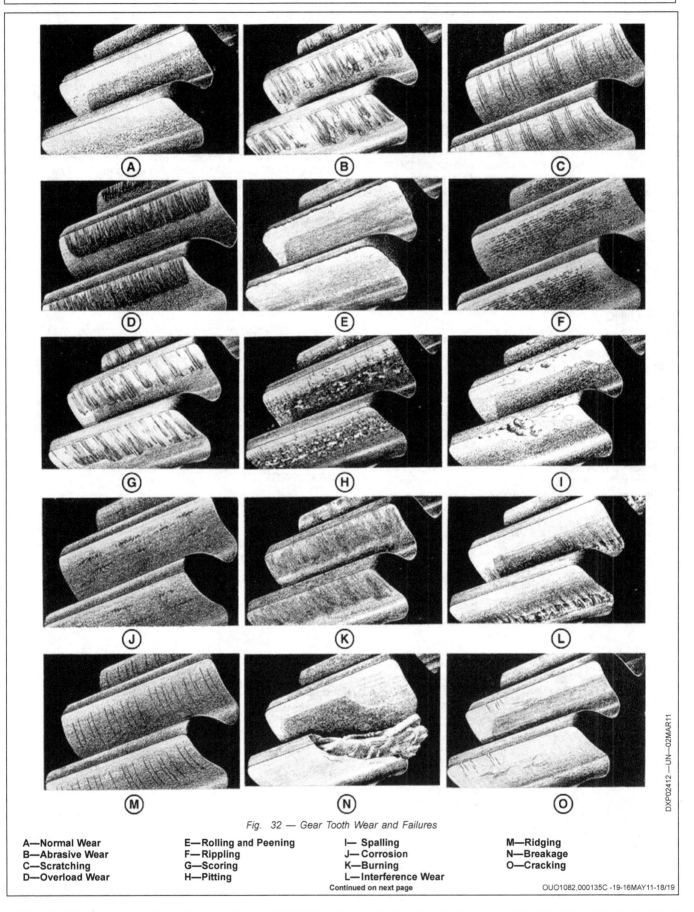

Fig. 32 — Gear Tooth Wear and Failures

A—Normal Wear
B—Abrasive Wear
C—Scratching
D—Overload Wear
E—Rolling and Peening
F—Rippling
G—Scoring
H—Pitting
I—Spalling
J—Corrosion
K—Burning
L—Interference Wear
M—Ridging
N—Breakage
O—Cracking

Continued on next page

POWER TRAINS—HOW THEY WORK

SUMMARY: GEARS

Gears are very basic to the function of power trains. In general, excessive backlash or endplay, or improper lubrication can lead to gear failures in the system.

Always be aware of the forces which are at work when gears are operating under load.

Alignment and fit are very important to ensure that each tooth absorbs its share of the load and that the gears are supported to resist thrust and twisting forces.

BEARINGS

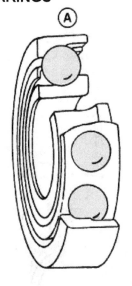 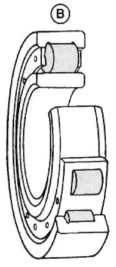 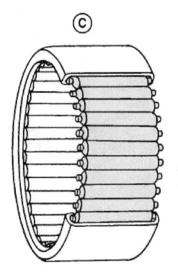

Fig. 33 — Three Types of Bearings

A—Ball Bearing   B—Roller Bearing   C—Needle Bearing

Bearings have two major jobs in a power train:

- Reduce friction
- Support a rotating shaft

We are mainly concerned with anti-friction bearings, those that give a rolling contact between mating surfaces (Fig. 33).

There are three main types:

- Ball
- Roller
- Needle

Continued on next page

POWER TRAINS—HOW THEY WORK

All these bearings are made of:

1. Two hardened-steel rings called races (Fig. 34).
2. Balls, rollers, or needles that roll between the two races.
3. Optional separators to space the rolling elements around the bearing.

On some bearings, the outer or inner race is omitted. Then the rolling elements are in direct contact with the shaft or other mounting (as in most needle bearings).

When two races are used, one race is normally pressed or fixed on a shaft or in a bore.

A—Outer Race
B—Ball
C—Shoulders
D—Bore Corner
E—Inner Ring Ball Race
F—Separator or Cage
G—Outer Ring Ball Race
H—Face
I—Bore
J—Inner Race

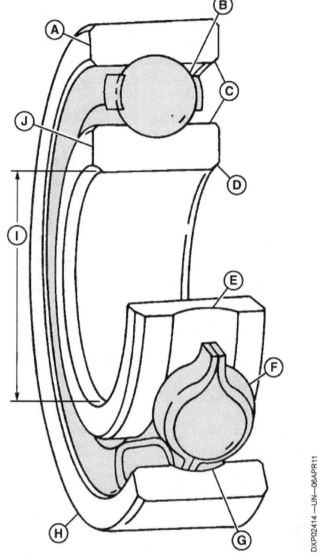

Fig. 34 — Basic Parts of a Bearing

Continued on next page

BEARING LOADS

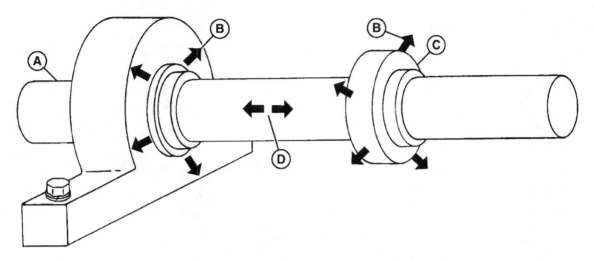

Fig. 35 — Load Forces Acting on Bearings

A—Shaft
B—Radial Forces
C—Bearing
D—End Thrust Forces

Bearing loads (Fig. 35) may be of two types:

1. Radial loads — forces perpendicular to the shaft.
2. Thrust loads — forces parallel to the shaft.

Many bearings must be able to carry both radial and thrust loads.

Now let's look at each type of bearing in more detail.

BALL BEARINGS

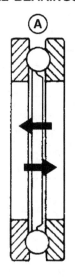 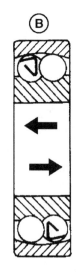 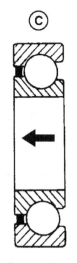 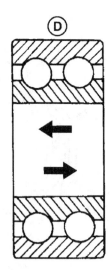 

Fig. 36 — Common Types of Ball Bearings

A—Ball Thrust
B—Self-Aligning
C—Radial-Thrust
D—Double Row
E—Single Row

Ball bearings support a shaft for radial forces as well as thrust forces (Fig. 36). The shaft must be aligned in the bearing bore, or the bearing will bind and quickly wear out.

To withstand various radial and thrust forces, the ball bearing types shown in Fig. 36 are widely used. (Thrust forces on each bearing are shown with arrows.) Ball bearings are also available in self-aligning types that compensate for shaft angles in relation to the bearing mount.

ROLLER BEARINGS

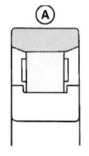 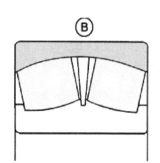  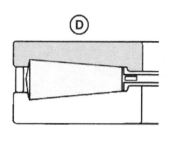

Fig. 37 — Common Types of Roller Bearings

A—Plain Roller
B—Self-Aligning Spherical Roller
C—Tapered Roller
D—Thrust Roller

Roller bearings are basically the same as ball bearings with the balls replaced by rollers, which can support greater loads. They come in various shapes, as shown in Fig. 37.

Straight roller bearings can handle heavy radial loads, but no thrust load.

Tapered roller bearings can withstand both radial and thrust loads. They must be installed in a way that the rollers are held in the proper position. This usually requires two matched bearings to prevent the shaft from moving in either direction, with careful adjustment of bearing positions.

Tapered roller bearings are commonly used where something (helical gears, bevel gears, external forces, etc.) places large thrust loads on the shaft.

Continued on next page

NEEDLE BEARINGS

Needle bearings (Fig. 38) are just specialized roller bearings. The rollers are much thinner, allowing many more to be used. The rollers are tightly packed, rather than separated by a cage.

Most needle bearings have no inner race. Instead, they roll directly on the shaft.

Each roller provides one line of contact to support the load, so a large number of needle rollers can carry far more radial load than fewer large rollers. But needle bearings cannot support any thrust load.

Needle bearings are used in compact locations where relatively high radial loads must be supported. This bearing is frequently located inside a gear that must run freely on a shaft or act as an idler. The length of the rollers and their tight packing give the gear good support and alignment. Planetary gears are usually supported by needle bearings.

While servicing, if there is doubt about the condition of a needle bearing, replace the bearing.

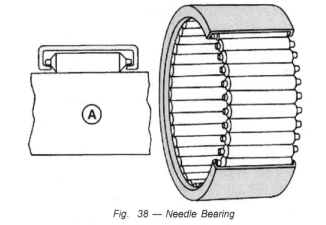

Fig. 38 — Needle Bearing

A—Shaft

For more information on bearings, refer to FOS-5406NC, Bearings and Seals.

ADJUSTING THE GEAR TRAIN

When a gear train is operated, reaction loads from gears, etc., are transmitted to the bearings and the various parts deflect.

For this reason, the gear train must normally be adjusted for the proper fit between parts.

Three kinds of adjustments are used:

- Backlash — clearance or "play" between gears in mesh
- Endplay — end-to-end movement in a gear shaft due to bearing clearances
- Preload — a load within the bearings set up by adjustment

Let's look at each one and then see how they all work together when adjusting an actual gear train.

BACKLASH IN GEARS

Too much backlash in gear trains is the result of either improper mesh between gears or lack of support in bearings.

The result of too much backlash can be broken gear teeth or bouncing of gears under impact forces.

Backlash is often adjusted to a specified reading on assembled gears (Fig. 39). Move ring gear in toward

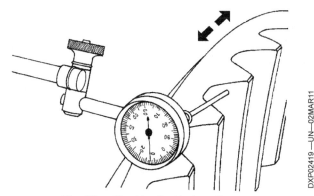

Fig. 39 — Checking Backlash on Gear

pinion to mesh gear teeth more deeply and reduce backlash, or away from pinion to increase (Fig. 41, Step 4).

The dial indicator is mounted so that it registers the full rotary movement of the ring gear shown.

To adjust the backlash reading, shims are often used (see following).

Continued on next page

ENDPLAY IN GEARS AND SHAFTS

Endplay refers to a measurable axial movement of a shaft or bearing. Endplay is measured with no load on the gear

To check endplay (Fig. 40), a dial indicator is mounted against the side of a gear or the end of a shaft. The gear or shaft is then pried in both directions and the readings noted. The difference between the two readings is the endplay.

Shims or adjusting nuts are widely used to adjust endplay.

The usual design objective is to have zero endplay under normal operating conditions.

Because steel expands with heat, we must make allowances. If expansion of the shaft will seat tapered roller bearings more deeply, we allow slight endplay when adjusting the bearings at room temperature.

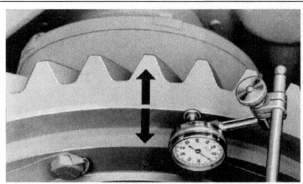

Fig. 40 — Checking Endplay on Gears and Shafts

Continued on next page

POWER TRAINS—HOW THEY WORK

PRELOADING OF GEAR TRAINS

Sometimes the opposite is needed. Tapered roller bearings are sometimes installed so that expansion of the shaft will increase endplay.

EXAMPLE: ADJUSTING RING GEAR AND PINION

Observe the pinion shaft bearings in Fig. 41. Because the spiral bevel gears exert a very large thrust load on the shaft, the bearing is positioned to support that load. The bearings on both ends must work as a set, so the other one has the large ends of the rollers turned outward.

Expansion of the shaft will increase the endplay. We want it to increase to zero at normal operating temperature, so we begin with a preload. When cold, the bearings are pulling on the ends of the shaft.

It is not possible to measure preload. What we do is first install enough shims to produce endplay. We measure the endplay, then remove enough shims to eliminate the endplay and add the specified preload.

Excessive endplay accelerates wear of gears and bearings, because they are not held in the proper position. Excessive preload is even worse, causing extreme overloading and rapid bearing failure.

Let's use all three of these adjustments — backlash, endplay, and preload — in adjusting the actual gear train.

A typical ring gear and pinion set is shown in Fig. 41. The pinion shaft at the top meshes with the ring gear at the bottom. The ring gear is supported in two quills, which bolt into the sides of the transmission housing.

When the gear set is assembled, it is adjusted in four steps:

1. Relation of pinion to ring gear is adjusted with shims under the lower shaft bearing.

2. The pinion shaft bearings are preloaded by using shims under the upper shaft bearing. (The shaft is first installed with extra shims, endplay is checked, and shims are removed to equal the endplay plus the specified preload.)

3. The ring gear bearings are preloaded by shims under the side quills. (Here again the ring gear is first installed with extra shims, endplay is checked, and shims are removed to equal the endplay plus the recommended preload for the bearings.) In this instance we use preload because the transmission housing will expand more than the ring gear assembly. We are aiming for zero endplay in the field.

4. Backlash between the ring gear and pinion is adjusted by transferring shims under the side quills. Backlash is checked at several points around the ring gear as shown in Fig. 39. To decrease the backlash, shims

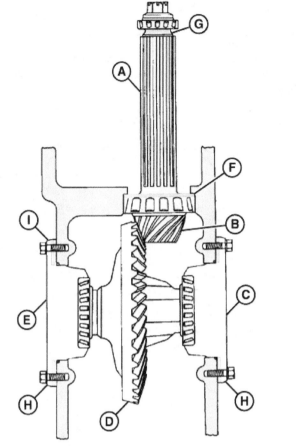

Fig. 41 — Adjusting Ring Gear and Pinion (A Typical Example)

A—Pinion Shaft
B—Pinion
C—Right Quill
D—Ring Gear
E—Left Quill
F—Step 1. Adjust relation of ring gear to pinion using shims here.
G—Step 2. Preload the pinion shaft bearings with shims here.
H—Step 3. Preload ring gear bearings with shims under two quills.
I—Step 4. Adjust backlash of ring gear and pinion by transferring shims between the quills.

are transferred from the left to the right quill. To increase backlash, shims are transferred from the right to the left quill. The same shims must be added to the opposite quill as removed from the other one, or the preload on the ring gear bearings will be changed.

This is a typical adjustment for a gear train, but there are many variations. Consult the machine technical manual for the exact steps and readings.

POWER TRAIN SAFETY

Manufacturers build many safety factors into their machines, but you must remember that not all potential hazards can be eliminated. The following safety rules are important to adhere to when working on or around power trains.

1. Always disengage the power and shut off the engine, remove the key, and wait for all parts to stop moving.

2. Keep guards and shields in place even when the machine is not operating (Fig. 42).

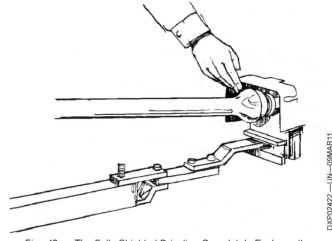

Fig. 42 — The Fully Shielded Driveline Completely Encloses the Machine-Connected Universal Joint and Coupler

3. Clean, lubricate, unplug, or hand feed a machine only when it is shut down and freewheeling parts have stopped moving (Fig. 43).

   A—Clutch

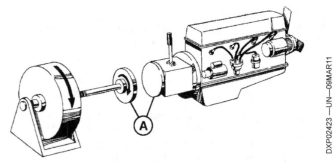

Fig. 43 — Freewheeling Parts Continue to Turn After the Power That Drives Them Is Disengaged — Inertia of Heavy Flywheel Causes It to Continue to Turn, Even After Power Is Disengaged

4. Clean rotating shafts. Rust, nicks, and dried mud or manure make them rough enough to catch clothing (Fig. 44).

   A—Nicks   C—Rust
   B—Dried Mud or Manure

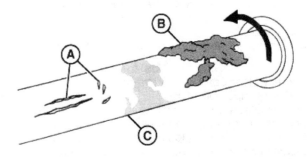

Fig. 44 — Even Seemingly Smooth Shafts Can Be Dangerous and Catch and Wrap Clothing

Continued on next page

5. Be aware of shaft ends that protrude beyond bearings; they are likely to wrap clothing (Fig. 45).

A—Leaning Against Protruding Shafts Is Dangerous

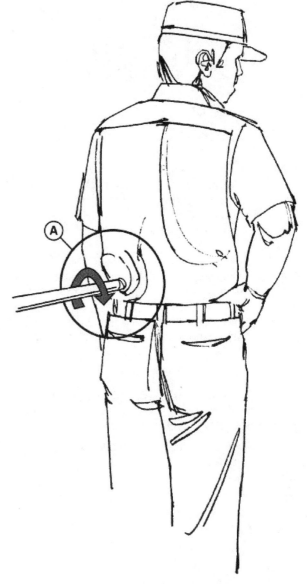

Fig. 45 — Shafts That Extend Much Beyond Bearings or Sprockets Can Be Dangerous

POWER TRAINS—HOW THEY WORK

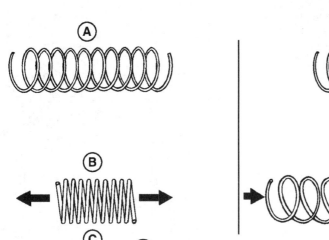
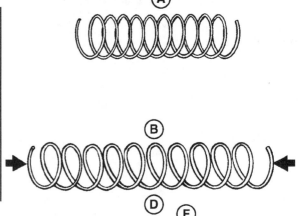

Fig. 46 — Energy Stored in Springs

A—No Energy, Spring Relaxed
B—Stored Energy
C—Spring Compressed — Energy Attempts to Push Ends Outward
D—Spring Stretched — Energy Attempts to Pull Ends Inward
E—Compression
F—Tension

6. When you remove any spring-loaded object, be sure you know what can happen. Know in which direction the spring will move, as well as all other components (Fig. 46).

7. Relieve pressure before disconnecting hydraulic lines and hoses. Keep hands and body away from fluids under high pressure (Fig. 47).

8. Follow instructions in operator's manuals.

Fig. 47 — Escaping Fluid Under Pressure Can Penetrate Skin Causing Serious Injury

POWER TRAINS—HOW THEY WORK

SUMMARY

NOTE: *An animated display of how the components in a basic Power Train works is available on the Instructor Art CD.*

We have seen how a basic power train works and have looked at the basic elements (Fig. 48).

In the rest of this book, we will look at various types of power trains in detail, starting with clutches in chapter 2.

A—Engine
B—Clutch
C—Transmission
D—Final Drive
E—Differential

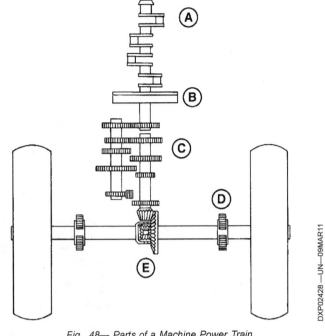

Fig. 48— *Parts of a Machine Power Train*

POWER TRAINS—HOW THEY WORK

Fig. 49 — Complete Power Train in Farm Tractor

TEST YOURSELF

QUESTIONS

1. Match each item on the left with the job it performs on the right.

   a. Clutch   1. Equalizes load for turning.
   b. Transmission   2. Connects and disconnects power.
   c. Differential   3. Selects speeds and direction.

2. Two gears are in mesh. Gear A has 12 teeth, while gear B has 24 teeth. When gear A has made one complete revolution, how far has gear B traveled?

   _____ 2 revolutions
   _____ 1/2 revolution
   _____ 4 revolutions

3. (True or False?) For speed reduction in low gears, a large gear drives a smaller gear.

4. Does reducing gear speeds increase or decrease the torque of the gear shaft?

5. (True or False?) When a machine turns, the outside drive wheel must turn faster.

6. What are the three basic ways of transmitting power?

7. Give an example of each type of transmitting power.

8. Name one advantage of helical gear teeth over straight spur gear teeth.

9. Name the two major jobs of bearings in a power train.

10. (Fill in the blank.) Bearings are often _____ to allow for expansion of the shaft.

11. (Fill in the blank.) Looseness in a power train from end to end is called _____.

12. (Fill in the blank.) Clearance between gear teeth in mesh is called _____.

CLUTCHES

INTRODUCTION

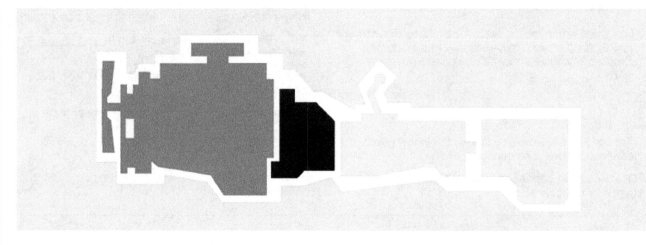

The clutch is a vital link in the power train. It connects and disconnects power between the source (engine) and the receiver (transmission).

Table 1 — Relation of Engine Clutch to Power Train

Engine = Source	Clutch = Link	Transmission = Receiver

The traction clutch, which provides power to drive the machine, may not be the only clutch. Other clutches engage and disengage PTO (power take-off) power to auxiliary equipment and MFWD (mechanical front wheel drive). Some tractors have front PTO. Industrial machines might use a clutch disconnect for auxiliary equipment such as a secondary pump.

Furthermore, a modern transmission may incorporate numerous clutches. Rather than using shifter forks to mechanically engage different gears, the gears are constantly in mesh and are engaged by hydraulic clutches.

A power shift transmission may include ten or more clutches and brakes. The only difference is that the clutch locks an element to a rotating shaft, whereas a brake locks it to a stationary housing.

CLUTCHES

TYPES OF CLUTCHES

Several types of clutches exist, including cone, band, expanding shoe, overrunning, centrifugal, fluid, and magnetic. However, almost all clutches used in agricultural and construction equipment are disk and plate type.

DISK AND PLATE CLUTCHES

The clutch may contain one disk or many. It may be dry or wet (cooled by oil). It may be engaged by spring force, a mechanical lever, or hydraulic pressure. It may or may not allow modulation for smooth engagement. The variety is impressive.

But the operating principle is always the same — to engage the clutch and transmit power, we squeeze one or more friction disks between smooth metal plates.

Fig. 1 shows a simple disk and plate clutch. It contains only one disk. It is a dry clutch, meaning it is not exposed to oil. It is spring engaged. It can only be released momentarily by pushing the clutch pedal. This is the type of clutch used on an automobile.

The clutch assembly is bolted to the engine flywheel. Strong springs squeeze the disk between the pressure plate and the flywheel. The disk is splined to the transmission input shaft.

To disengage the clutch, we push the clutch release bearing (often called a throw-out bearing) against the clutch release levers. The levers pivot near the outer ends, and pull the pressure plate back against the spring force. Actual movement of the pressure plate is very slight, but enough to release the disk from its vise-like grip. This enables you to stop the machine or to shift gears.

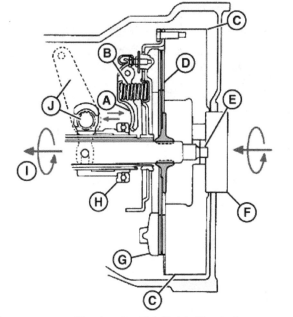

Fig. 1 — Dry Disk Clutch (Standard)

A—Clutch Release Lever
B—Clutch Spring
C—Engine Flywheel
D—Clutch Driven Disk
E—Transmission Input Shaft
F—Engine Crankshaft
G—Clutch Pressure Plate
H—Clutch Release Bearing
I—To Transmission
J—Clutch Release Linkage

Fig. 2 shows one example of this simple disk and plate clutch. This one uses three formed steel sections for the clutch release levers. This is a fairly light duty clutch.

A—Clutch Disk
B—Pressure Plate

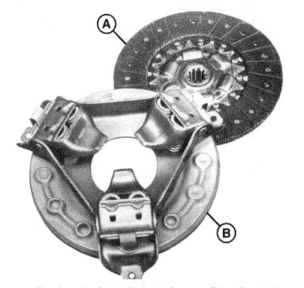

Fig. 2 — Dry Clutch Disk and Pressure Plate (Standard)

CLUTCHES

Fig. 3 shows a somewhat heavier clutch with three adjustable fingers to contact the release bearing.

Still other pressure plates use a slotted spring steel ring with a full circle of contact points instead of individual release levers. All of them perform the same function — to pull the pressure plate back when the release bearing is pushed forward.

A—To Clutch Pedal
B—Engine Flywheel
C—Clutch Disk
D—Pressure Plate
E—Clutch Shaft
F—Clutch Springs
G—Clutch Cover
H—Clutch Release Levers
I—Clutch Release Bearing with Carrier
J—To Transmission
K—Clutch Fork

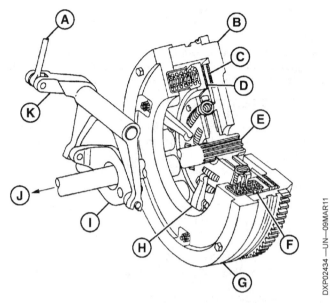

Fig. 3 — Heavy Duty Disk Clutch (Engaged)

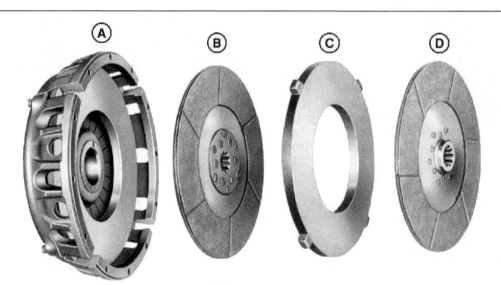

Fig. 4 — Parts of a Heavy Duty Disk Clutch, Which Uses Two Disks for Increased Capacity

A—Pressure Plate
B—First Clutch Disk
C—Intermediate Plate
D—Second Clutch Disk

Fig. 4 shows an even heavier clutch. It uses two disks for higher capacity. The intermediate plate has tangs that fit into slots in the pressure plate assembly so it rotates with the flywheel. The pressure plate squeezes each friction disk between the two smooth metal surfaces. This type of clutch is common in large trucks. By adding one disk and one plate, the capacity is doubled.

CLUTCHES

DESIGN CONSIDERATIONS

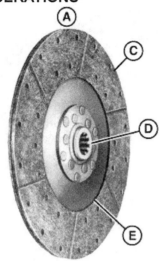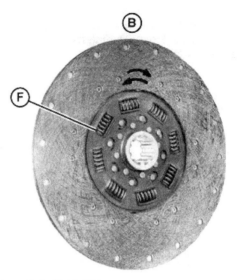

Fig. 5 — Heavy Duty Clutch Disks

A—Rigid Disk
B—Flexible Disk
C—Facing
D—Hub
E—Driven Plate
F—Torsional Damper Springs in Hub

CAPACITY

The amount of torque a clutch can transmit depends on:

- Number of disks and plates
- Diameter of disks and plates
- Force exerted by the pressure plate
- Coefficient of friction

DISK

The disk facing can be made of organic fiber or ceramic material. Sometimes metallic flakes are embedded for longer life. Asbestos performed well in clutch and brake facings, but its use was discontinued after health concerns were discovered.

The facing may be a continuous ring, or it may be in sections similar to disk brake pads, often called a "button clutch." The surface may be smooth or grooved. Each variation has loyal followers, and they'll probably never agree on which is the best.

Fig. 5 shows another feature found only in "flexible" clutch disks. Torsional damper springs provide a cushion. This protects the transmission from torque spikes produced by the engine, reducing backlash chatter in the gears. It also protects the engine from torque feedback, perhaps from impacts or torsional pulsations in a PTO driveline.

Looking back to the cutaway view in Fig. 3, you can see how torsional damper springs work. The disk hub is sandwiched between two layers connected to the disk facing, with slight movement allowed. The damper springs hold the hub in position, letting it absorb spikes by compressing the springs.

The clutch disk is actually the heart of the clutch. Check for high and low spots of wear on the friction facings. On riveted linings, use the level of material above the rivet head as an indicator of wear. Replace the linings if the rivet heads are flush or just below the outer surface.

Replace the disk if it appears glazed or cracked. Check any cushion springs and replace if damaged.

Make one last check of the clutch disk: Slide the disk onto the input shaft of the transmission. It should slide easily without restriction. If the disk rocks back and forth or from side to side, replace it. This may mean that the splines are badly worn. Try a new disk on the clutch shaft. If you get the same motion, the input shaft is worn and it will have to be replaced.

A good indication of clutch alignment is the clutch disk hub. If there is uniform wear along the splines, this means the clutch is aligned. If there is excessive wear on the front of the spline, this means the clutch is misaligned.

PRESSURE PLATE ASSEMBLY

As you can see by comparing Figs. 2–4, pressure plate designs vary widely — thick or thin, heavy cast iron or light stamped steel. They may use a few coil springs, many coil springs, or one large spring washer to apply force. The release levers come in all shapes and sizes.

Regardless of design details, the principles are always the same. The assembly is bolted to the flywheel. Very strong springs squeeze a disk between the pressure plate and the flywheel to engage the clutch. The release levers act as pivots to pull the pressure plate back a small distance when the release bearing pushes them forward.

Continued on next page

CLUTCHES

Fig. 6 shows a pressure plate being checked for warpage. The surface can be machined if warped or scored, but only within specified limits. Removing too much material would reduce spring force and allow the clutch to slip too easily.

A—Straight Edge
B—Feeler Gauge
C—Clutch Plate

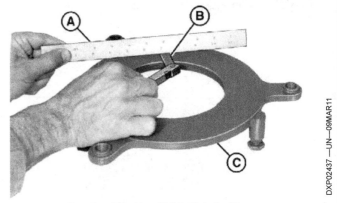

Fig. 6 — Checking Clutch Plate for Warpage

Continued on next page

CLUTCHES

Fig. 7 shows a tool used to check pressure plate springs. All springs must be uniform and must exert a specified force when compressed to a specified length.

TRANSMISSION INPUT SHAFT

Most input shafts have a smaller shaft or pilot that projects from the front end. This pilot rides in the pilot bearing in the engine crankshaft flange.

The splined area is provided for the clutch disk. It must allow the clutch disk to move laterally along the spline but must not allow a rotary rocking motion.

A machined area between the transmission front cover and the spline mounts the release bearing carrier.

A thrust-type ball bearing is the most common release bearing. It rests on this area of the shaft when the clutch is engaged.

Check the release bearing to see if it rotates freely. If in doubt about the wear or lack of lubrication, replace this bearing.

FLYWHEEL

A flywheel performs several engine-related functions:

- It acts as a balancer for the engine, leveling out power impulses.
- It provides a ring gear for the engine starting motor to contact.

As a power train component, it provides a mounting surface and a friction surface for the clutch.

The end of the crankshaft where the flywheel is installed usually has a small hole in the center with a pilot bushing. This is a guide for the tip of the clutch shaft.

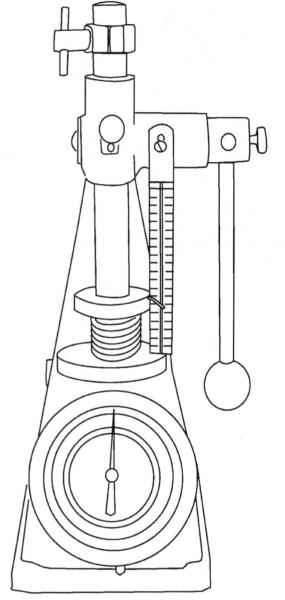

Fig. 7 — Checking Clutch Springs

CLUTCHES

Fig. 8 shows a flywheel with a clutch removed. This friction surface can also be machined if necessary, but the dimension between the mounting surface and friction surface is critical. A few flywheels have replaceable wear plates to reduce the cost if replacement is necessary.

There is a specification for surface finish for both the flywheel and pressure plate. People sometimes assume that the surfaces should be as smooth as possible, but we do not want a mirror finish. An excessively smooth surface may increase slippage or cause the disk to stick when you're trying to disengage it. An excessively rough surface causes rapid disk wear.

Look for heat checks on the flywheel surface. If the check markings are excessive, the flywheel must be replaced.

If the pilot bearing for the transmission input shaft appears out-of-round or badly worn, it should be replaced. Score marks on the pilot shaft indicate a bearing failure.

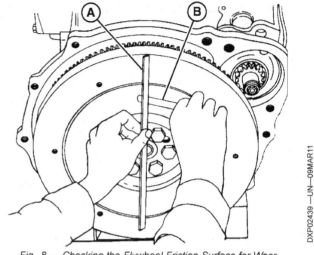

Fig. 8 — Checking the Flywheel Friction Surface for Wear

A—Straight Edge **B—Feeler Gauge**

CLUTCHES

CLUTCH LINKAGE

STANDARD MECHANICAL LINKAGE

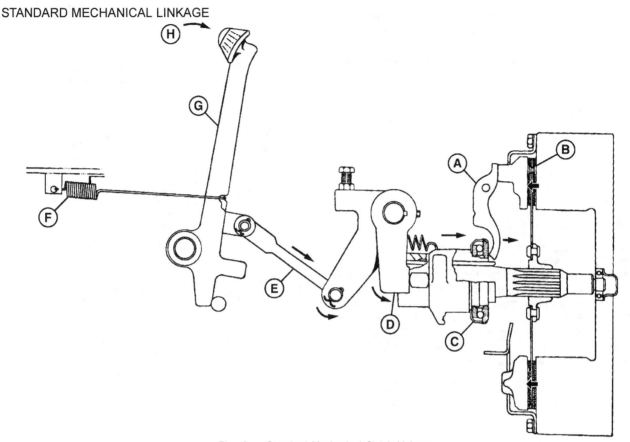

Fig. 9 — Standard Mechanical Clutch Linkage

A—Clutch Release Lever
B—Clutch (Disengaged)
C—Clutch Release Bearing
D—Fork
E—Linkage
F—Return Spring
G—Clutch Pedal
H—Disengaging Movement

Fig. 9 shows an example of a simplified clutch linkage. When the clutch pedal is pushed down, a fork inside the clutch housing pushes the release bearing forward. This pivots the release levers to pull the pressure plate back.

Usually (but not always) the linkage is designed to pull the release bearing away from the release levers when the pedal is all the way up. This ensures that the clutch is fully engaged when the pedal is up. It also extends the life of the bearing, because the only time it turns is when the pedal is pushed in. This is why "riding the clutch" can cause premature failure of the release bearing.

Pedal "free travel" is an important adjustment. It is a measurement of how far the pedal moves before the release bearing makes contact with the release levers.

SERVICING MECHANICAL CLUTCH CONTROLS

When repairing the clutch, always examine the linkage.

Worn bushings, bent rods, broken springs, and damaged cotter pins can result in excessive force being required to operate the clutch.

Clutch pedal free travel is often used as a guide to conditions inside the clutch.

The actual clutch pedal free travel is more easily determined if checked by hand and should be the clearance between the clutch release yoke fingers and the release bearing housing.

Clutch pedal free travel varies; set it according to technical manual instructions.

Check the release yoke movement to verify that free pedal travel is actually the release bearing clearance, and not lost motion in linkage due to worn clevis eyes and pins, or worn bushings and shafts at the clutch pedal or linkage arms.

Continued on next page

CLUTCHES

HYDRAULIC CONTROLS

Many machines use a hydraulic release mechanism instead of a solid linkage (Fig. 10). The clutch pedal is connected to a master cylinder. Pushing the pedal sends oil through flexible pressure hose or metal tubing to a slave cylinder, which releases the clutch.

Hydraulic linkage typically is not designed to pull the release bearing away from the release levers. Instead, it uses a larger bearing, which is meant to run full-time.

Hydraulic linkage has an equivalent of clutch pedal free travel. It is important to make sure trapped oil isn't partially releasing the clutch. The pedal should move a specified distance before starting to send pressurized oil to the slave cylinder.

SERVICING HYDRAULIC CONTROLS

- Adjust the clutch pedal free travel according to the technical manual.
- Watch for leaking seals, low fluid level, broken or kinked lines, and leaking pistons in the cylinders.
- Always bleed the air from the system after repair or replacement of parts. In most designs, a special screw is provided to allow bleeding the air from the system.
- Check the fluid level in the reservoir and add oil if necessary after bleeding the system.

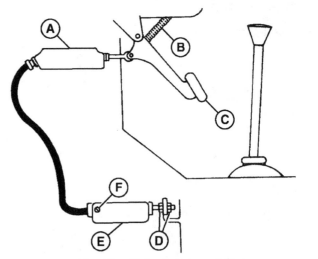

Fig. 10 — Hydraulic Linkage Controls

A—Master Cylinder
B—Return Spring
C—Clutch Pedal
D—Adjusting Nuts
E—Slave Cylinder
F—Bleed Screw

- Adjust the slave cylinder rod to set the clutch throw-out bearing at the proper position.

Continued on next page

CLUTCHES

DUAL CLUTCH ASSEMBLIES

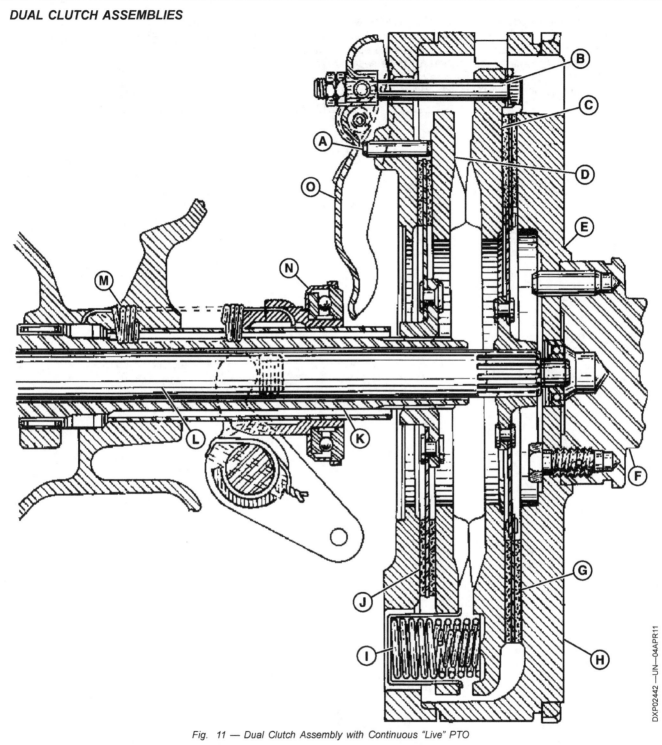

Fig. 11 — Dual Clutch Assembly with Continuous "Live" PTO

A—PTO Clutch Release Pin
B—Operating Bolt
C—Transmission Clutch Plate
D—PTO Clutch Plate
E—Flywheel
F—Engine Crankshaft
G—Transmission Clutch Disk
H—Clutch Cover
I—Springs
J—PTO Clutch Disk
K—PTO Clutch Shaft
L—Transmission Clutch Shaft
M—Return Spring
N—Throw-Out Bearing
O—Operating Lever

All of the examples so far have been single clutches. Even if it used two disks for higher capacity, the clutch controlled only one function.

Now let's look at an example that combines two separate clutches into one assembly (Fig. 11). One is the traction clutch, like those previously discussed. The other is the PTO clutch.

Continued on next page

CLUTCHES

This version is called "continuous running" or "live" PTO. It uses one clutch pedal and one release bearing, but there are two sets of release levers and two pressure plates. The release bearing contacts the traction clutch release levers first. Pushing the pedal farther, the release bearing then makes contact with the PTO clutch release levers.

The advantage is that the PTO can be started before the tractor moves. The machine can also be stopped without disengaging the PTO.

Continued on next page

CLUTCHES

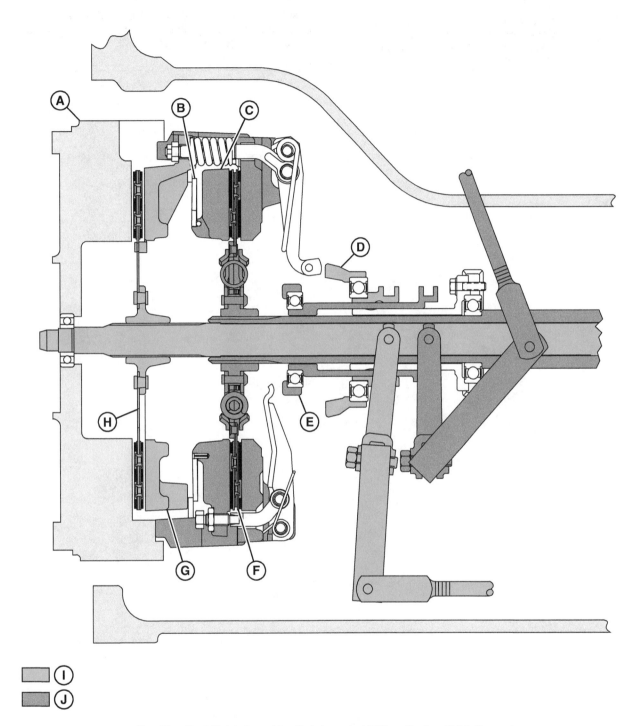

Fig. 12 — Dual Clutch Assembly with Independent PTO — Traction Clutch Engaged

A—Flywheel
B—Spring Washer
C—Traction Clutch Pressure Plate
D—PTO Clutch Engagement Bearing
E—Traction Clutch Release Bearing
F—Traction Clutch Disk
G—PTO Clutch Pressure Plate
H—PTO Clutch Disk
I—PTO Clutch Components
J—Traction Clutch Components

Better yet is the "independent" PTO as shown in Fig. 12. There are still two clutches in one assembly, but now each is controlled by separate linkage. The clutch pedal uses one bearing to release the traction clutch. A separate lever controls another bearing to operate the PTO clutch.

Continued on next page

CLUTCHES

The traction clutch is the rear half of the assembly. Notice that the movable pressure plate is in front of the disk. A spring washer, held in place by a snap ring around the outside, provides a large force against the pressure plate to squeeze the disk against a stationary plate behind it. The traction clutch is engaged.

Continued on next page

CLUTCHES

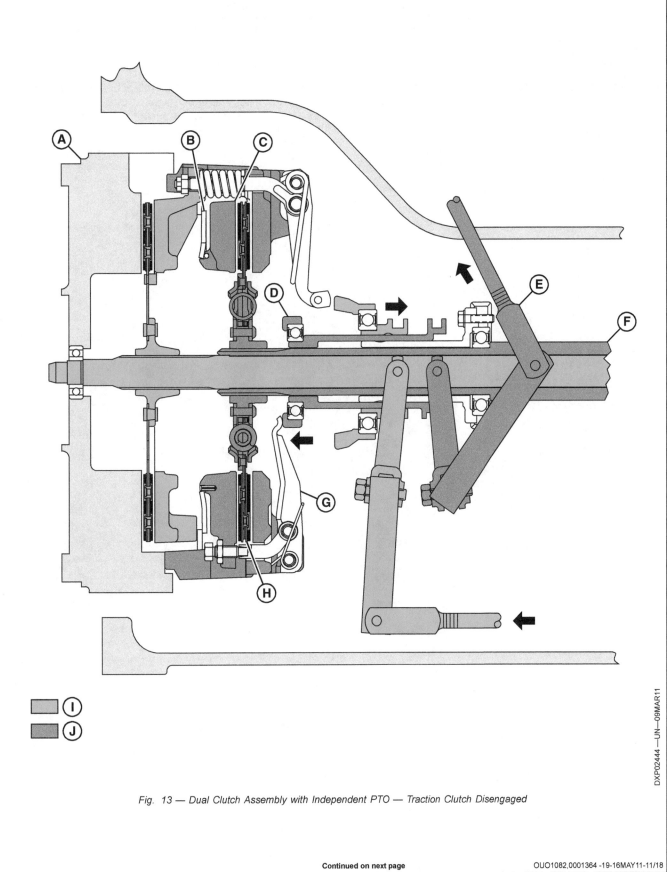

Fig. 13 — Dual Clutch Assembly with Independent PTO — Traction Clutch Disengaged

CLUTCHES

A—Flywheel
B—Spring Washer
C—Traction Clutch Pressure Plate
D—Traction Clutch Release Bearing
E—Clutch Pedal Linkage
F—Traction Clutch Shaft
G—Traction Clutch Release Levers
H—Traction Clutch Disk
I—PTO Clutch Components
J—Traction Clutch Components

Now watch what happens when the clutch pedal is pushed down (Fig. 13). The inner bearing slides forward and contacts one set of clutch release levers. The diagram shows only one lever for the traction clutch and one for the PTO clutch, but there are three of each. The levers pivot at the outer end. They push against three adjustable pins threaded into the movable pressure plate, compressing the spring washer and releasing the traction clutch.

Continued on next page

CLUTCHES

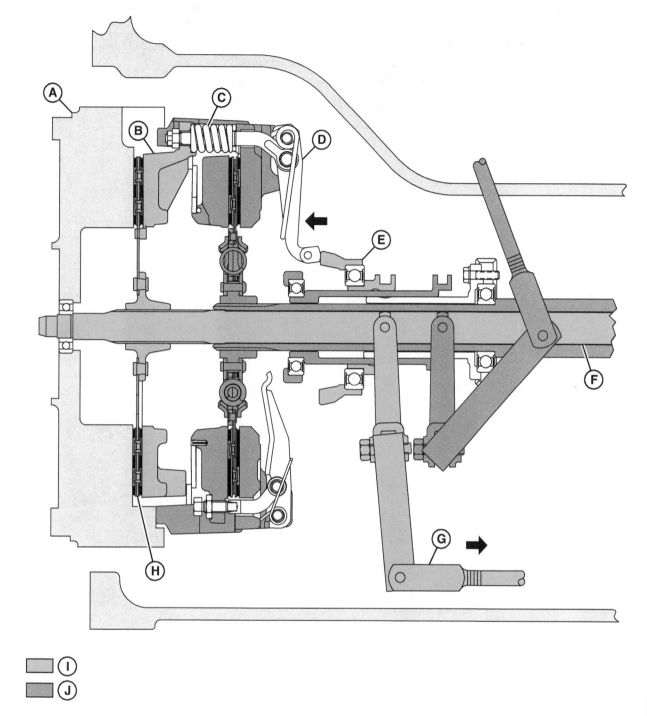

Fig. 14 — Dual Clutch Assembly with Independent PTO — PTO Clutch Engaged

A—Flywheel
B—PTO Clutch Pressure Plate
C—PTO Clutch Engagement Springs
D—PTO Clutch Engagement Levers
E—PTO Clutch Engagement Bearing
F—PTO Clutch Shaft
G—PTO Control Lever Linkage
H—PTO Clutch Disk
I—PTO Clutch Components
J—Traction Clutch Components

This example has another twist. Unlike all previous clutches, we push the bearing against the levers on the pressure plate to engage the PTO clutch, instead of to release it. The control lever goes "over center" to hold it in the engaged position.

Continued on next page

CLUTCHES

The two previous diagrams showed the PTO clutch disengaged. Small springs push the levers back and pull the pressure plate away from the disk.

Fig. 14 shows the PTO clutch engaged. The larger bearing pushes against the three PTO clutch engagement levers. The levers pivot at the outer end. They push the three long pins, which push the other movable pressure plate forward and squeeze the PTO clutch disk against the flywheel. Note the heavy coil spring on each pin. The pins push against the springs, and the springs push against the pressure plate. This provides a bit of cushion in the linkage.

Notice the concentric shafts (Fig. 15).

Continued on next page

CLUTCHES

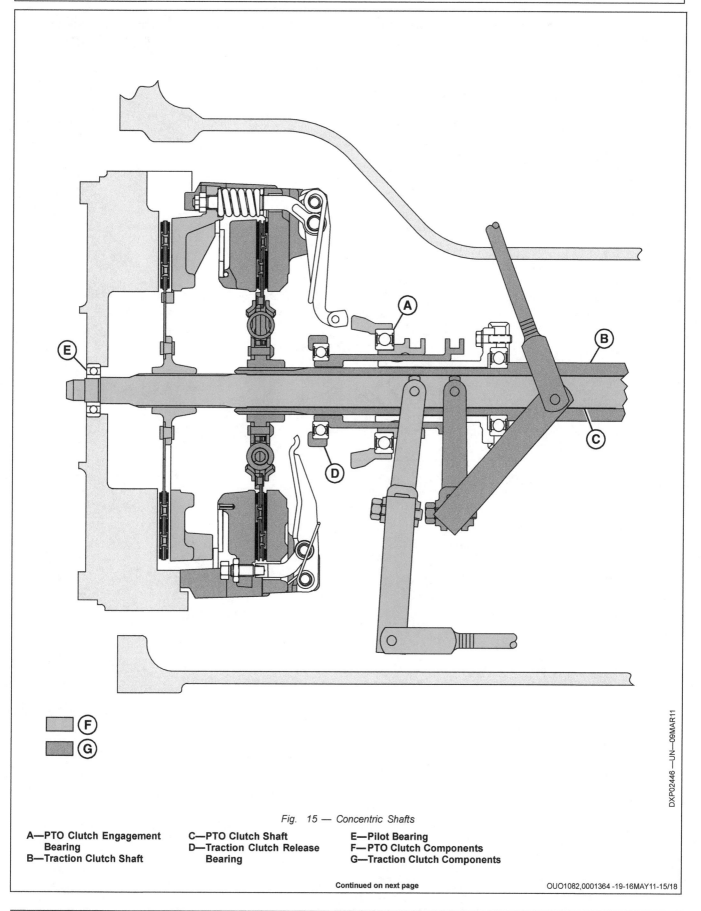

Fig. 15 — Concentric Shafts

A—PTO Clutch Engagement Bearing
B—Traction Clutch Shaft
C—PTO Clutch Shaft
D—Traction Clutch Release Bearing
E—Pilot Bearing
F—PTO Clutch Components
G—Traction Clutch Components

Continued on next page

CLUTCHES

- The PTO clutch shaft is in the center. It is a solid shaft. The front tip fits into a pilot bearing in the flywheel to keep everything aligned.
- The traction clutch shaft is hollow. It fits over the outside of the PTO clutch shaft.
- The traction clutch release bearing is on a tube that fits over the outside of the traction clutch shaft. It does not rotate, but slides back and forth, controlled by a fork.
- The PTO clutch engagement bearing is on a tube that fits over the outside of the previous tube. It also slides back and forth, controlled by a second fork.

CLUTCHES

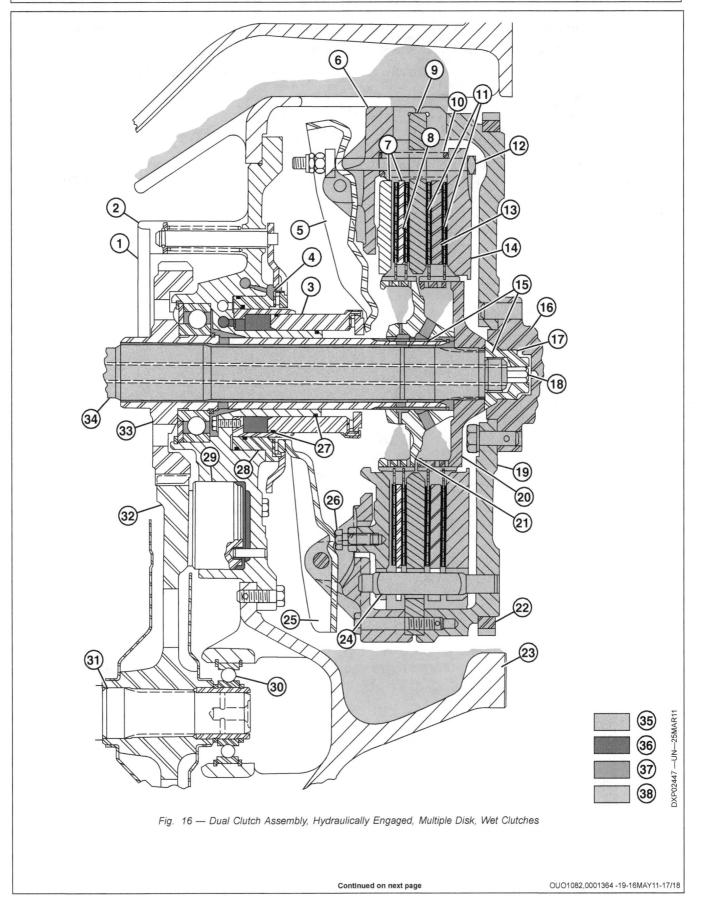

Fig. 16 — Dual Clutch Assembly, Hydraulically Engaged, Multiple Disk, Wet Clutches

Continued on next page

CLUTCHES

1— 2-Speed Brake Housing
2— Operating Bearing
3— Clutch Operation Piston
4— Needle Bearings
5— Clutch Operating Lever
6— Clutch Cover
7— PTO Clutch Disks
8— PTO Separator Plate
9— Clutch Backing Plate
10— Lever Return Spring
11— Transmission Clutch Disks
12— Clutch Operating Bolt
13— Transmission Separator Plate
14— Clutch Pressure Plate
15— Bushings
16— Crankshaft
17— Pump Drive Adapter
18— Hex Drive Shaft
19— Flywheel
20— Transmission Clutch Hub
21— PTO Clutch Hub
22— Ring Gear
23— Clutch Housing
24— Drive Pin
25— PTO Operating Lever
26— PTO Adjusting Screw
27— Seals
28— PTO Operating Piston
29— PTO Brake Piston
30— Bearing Assembly
31— PTO Shaft
32— PTO Drive Gear/Oil Shield
33— PTO Gear and Shaft
34— Clutch Shaft
35— Power Flow
36— Transmission Control Oil
37— Lubrication Oil
38— Pressure Free Oil

Let's look at one more variation in Fig. 16. It's another dual clutch assembly mounted on the flywheel, but different from any of the others:

- Both clutches are engaged by pushing on the levers.
- Both use internal hydraulic cylinders instead of mechanical linkage to push the levers.

- Each clutch has two disks splined to a center hub, separated by a steel plate tanged to the outer housing.
- They are "wet" clutches. Continuous oil spray cools and lubricates all parts.

CLUTCHES

CLUTCHES IN OTHER LOCATIONS

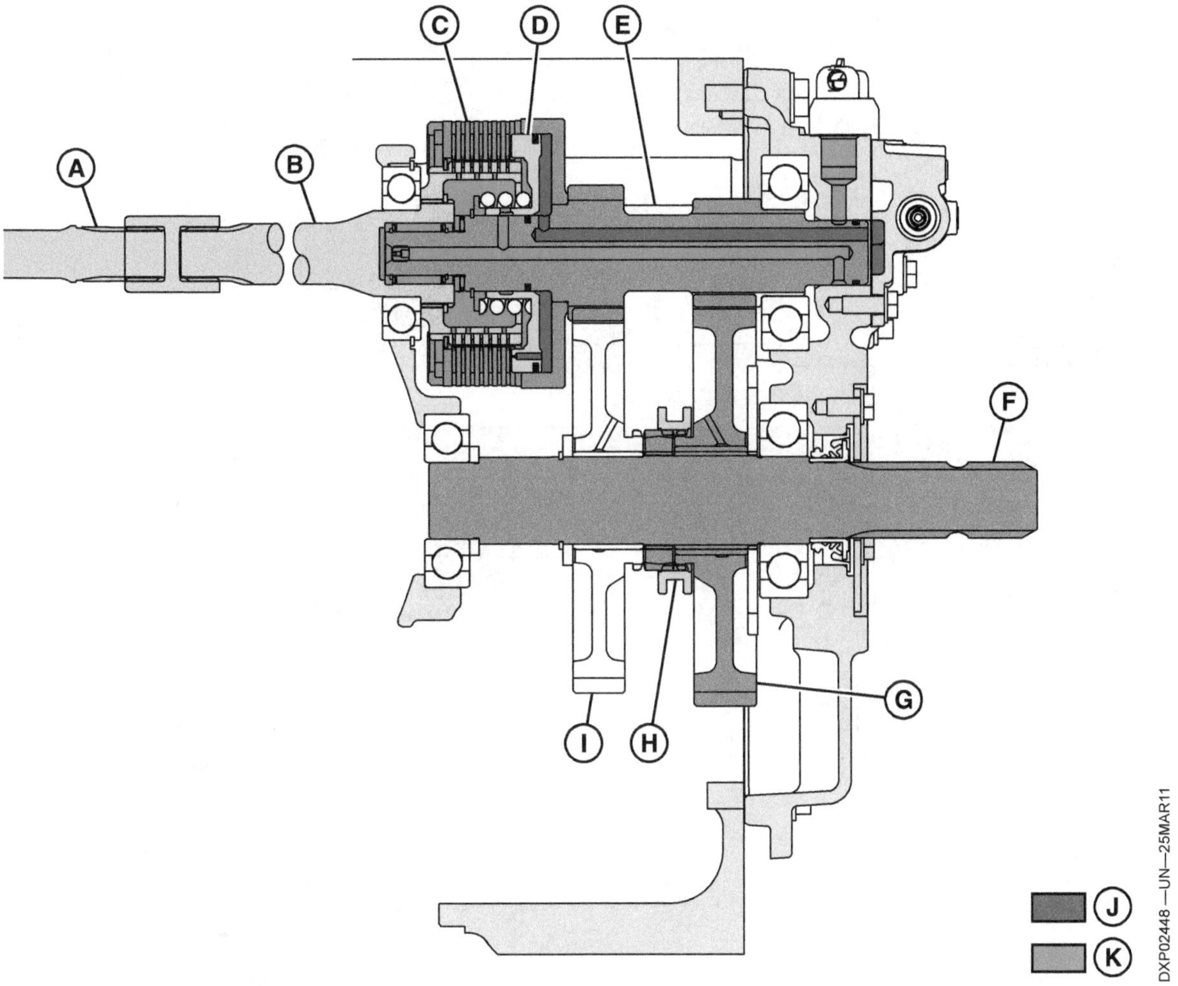

Fig. 17 — Electrohydraulic PTO Clutch

A—PTO Shaft
B—Intermediate Shaft
C—PTO Clutch Pack
D—Piston
E—PTO Pinion Shaft
F—PTO Output Shaft
G—PTO Drive Gear for Standard 540 Mode
H—PTO Shift Collar
I—PTO Drive Gear for Economy 540 Mode
J—Pressure Oil
K—Lubrication Oil

All the clutches we've seen so far are mounted on the engine flywheel. That's a convenient starting point, but clutches can be anywhere.

For instance, here's a PTO clutch (Fig. 17). It's at the rear of a tractor. It's a wet clutch with seven disks, enabling the small diameter to carry the full engine power. The disks are splined to a hub on the input shaft, which turns all the time. The separator plates are tanged to a drum on the pinion shaft.

To engage the PTO, the operator just flips a switch. An electrohydraulic valve on the back of the housing sends pressurized oil through a drilled passage in the pinion shaft. This pushes a piston inside the clutch drum to squeeze the disks and plates. When the engagement pressure is released, a coil spring inside the clutch pushes the piston back. It's as simple as that.

Continued on next page

CLUTCHES

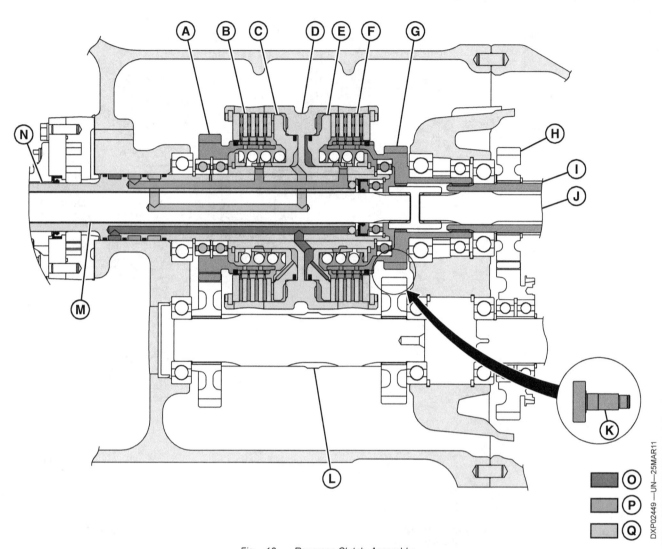

Fig. 18 — Reverser Clutch Assembly

A—Reverse Drive Gear
B—Reverse Drive Plates and Disks
C—Reverse Piston
D—Clutch Drum
E—Forward Piston
F—Forward Plates and Disks
G—Forward Drive Gear
H—Transmission Second Speed Drive Gear
I—Transmission Top Shaft
J—PTO Shaft
K—Reverse Idler Gear
L—Countershaft Speed Sensor Teeth
M—PTO Drive Shaft
N—Traction Drive Shaft
O—Pressure Oil
P—Lubrication Oil
Q—Pressure Free Oil

Fig. 18 shows a hydraulic reverser assembly for a transmission. It contains two wet clutches similar to the one described previously. Besides determining which direction the machine will move, it also replaces the type of traction clutch previously discussed.

The traction drive shaft is driven by the engine flywheel; it turns all the time. It is splined to a common drum for both clutches. Each clutch has four disks splined to a hub that is attached to the forward and reverse drive gears. Each clutch has an internal piston for hydraulic engagement and a coil spring for disengagement. Oil is directed to either clutch (never both) to lock its gear to the input shaft. Either end will turn the countershaft, but one end uses an idler gear to reverse the direction of rotation.

The clutch pedal operates a valve that controls pressure to the reverser. A lever beside the steering wheel determines whether the forward clutch or the reverse clutch receives the pressure.

Continued on next page

OUO1082,0001365 -19-16MAY11-2/8

CLUTCHES

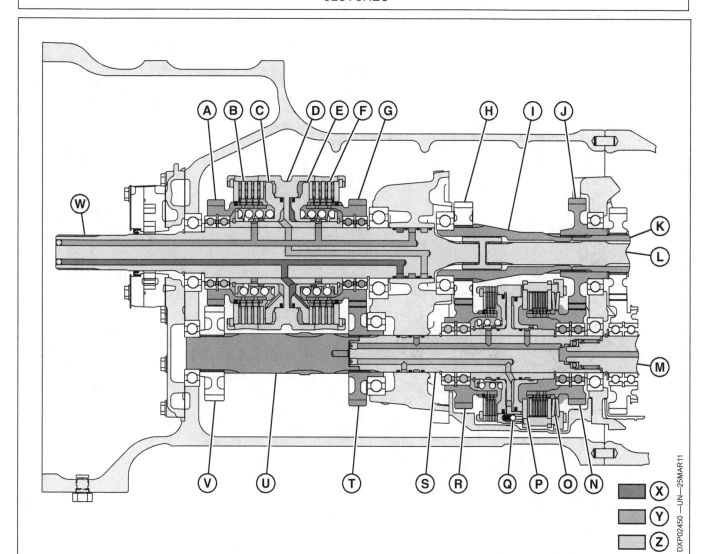

Fig. 19 — Hi-Lo Clutch Assembly

A—Reverse Drive Gear
B—Reverse Drive Plates and Disks
C—Reverse Piston
D—Clutch Drum
E—Forward Piston
F—Forward Plates and Disks
G—Forward Drive Gear
H—High Speed Driven Gear
I—Connect Shaft
J—Low Speed Driven Gear
K—Transmission Top Shaft
L—PTO Shaft
M—Transmission Bottom Shaft
N—Low Speed Drive Gear
O—Low Side Engagement Springs
P—Clutch Plate
Q—Bleeder
R—High Speed Drive Gear
S—Hi-Lo Clutch Shaft
T—Forward Driven Gear
U—Countershaft
V—Reverse Driven Gear
W—Traction Drive Shaft
X—Pressure Oil
Y—Lubrication Oil
Z—Pressure Free Oil

Stepping back for a wider view, we can see the reverser is connected to a hi-lo assembly (Fig. 19). The hi-lo function is almost the same. The shaft is splined to a common drum for the hi and lo clutches. Both are multi-disk wet clutches. Each clutch has disks splined to a hub that is attached to a gear. Neither gear uses a reverse idler, so this assembly provides two speeds in the same direction.

There is one detail difference between reverser and hi-lo clutches, The lo clutch is spring-engaged. Note the two spring washers that squeeze the stack of disks and plates together. The hi-lo is always in lo when the engine is started. It will automatically downshift to lo if control pressure drops enough to risk slipping the clutch.

When the control lever is moved to hi, a valve sends pressurized oil to both the hi and lo clutch pistons. This engages the hi clutch in the same manner as shown previously for the PTO and reverser clutches, but it releases the lo clutch by compressing the spring washers. The lo clutch piston doesn't push against the disks and plates; it pushes the pins at the outer edge, which moves a plate back to release the disks and plates.

Continued on next page

CLUTCHES

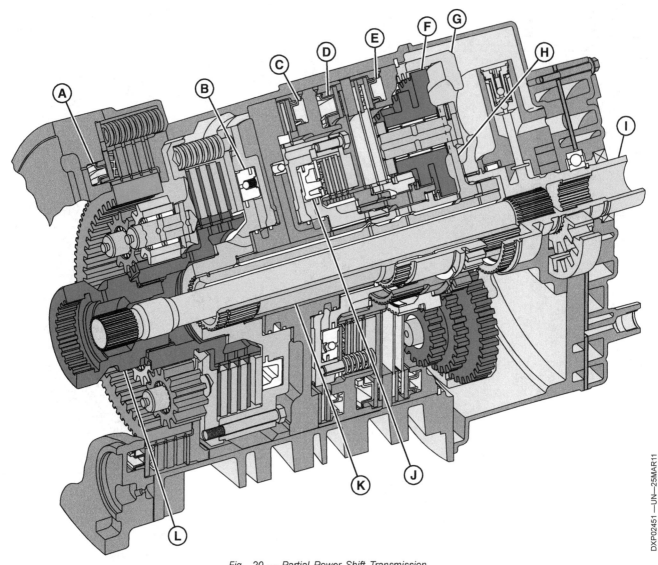

Fig. 20 — Partial Power Shift Transmission

A—Reverse Brake
B—Forward Clutch
C—Third Speed Brake
D—Second Speed Brake
E—First Speed Brake
F—Planet Pinion Gear
G—Ring Gear
H—Planet Pinion Carrier
I—Input Shaft
J—Fourth Speed Clutch
K—Forward Clutch Drive Shaft
L—Output Shaft

Other transmissions use even more clutches. Fig. 20 shows a "partial power shift" assembly, which provides four forward and four reverse speeds. As with the previous example, this is used in conjunction with a mechanical gearbox. You have all these selections within each range.

This includes a forward clutch, reverse brake, first speed brake, second speed brake, third speed brake, and fourth speed clutch. Remember, the only difference between a clutch and a brake is its usage — whether it is engaged to turn an element or prevent it from turning.

The engine flywheel turns the input shaft, which is splined to the ring gear for the triple compound planetary pinions. Three brakes and a clutch provide three reductions and direct drive for the planet pinion carrier, which is splined to the forward clutch drive shaft. Then the forward clutch or reverse planetary engages the output shaft.

Continued on next page

CLUTCHES

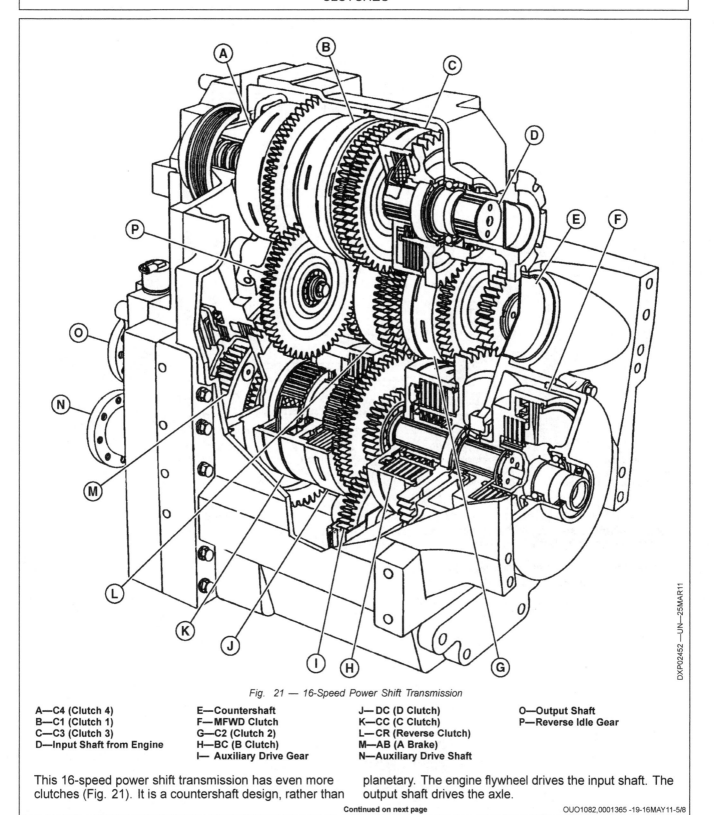

Fig. 21 — 16-Speed Power Shift Transmission

A—C4 (Clutch 4)
B—C1 (Clutch 1)
C—C3 (Clutch 3)
D—Input Shaft from Engine
E—Countershaft
F—MFWD Clutch
G—C2 (Clutch 2)
H—BC (B Clutch)
I—Auxiliary Drive Gear
J—DC (D Clutch)
K—CC (C Clutch)
L—CR (Reverse Clutch)
M—AB (A Brake)
N—Auxiliary Drive Shaft
O—Output Shaft
P—Reverse Idle Gear

This 16-speed power shift transmission has even more clutches (Fig. 21). It is a countershaft design, rather than planetary. The engine flywheel drives the input shaft. The output shaft drives the axle.

Continued on next page

CLUTCHES

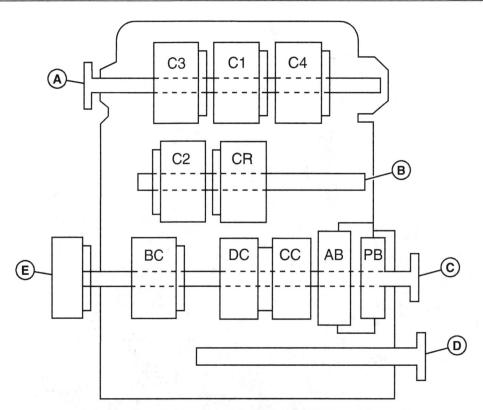

Fig. 22 — Clutches in a 16-Speed Power Shift Transmission

A—Input Shaft
B—Countershaft
C—Output Shaft
D—Auxiliary Drive Shaft
E—MFWD
AB—A Brake
BC—B Clutch
C1—Clutch 1
C2—Clutch 2
C3—Clutch 3
C4—Clutch 4
CC—C Clutch
CR—Reverse Clutch
DC—D Clutch
PB—Park Brake

Inside the transmission, eight clutches and one brake provide the speed selections, plus a park brake (Fig. 22). The PTO clutch and MFWD clutch are located elsewhere.

Leaving power flows for another chapter, each speed uses one of the upper elements (C1, C2, C3, C4, or CR) and one of the lower elements (AB, BC, CC, or DC). "PB" is the park brake. This produces 16 speeds forward and 4 reverse.

Continued on next page

CLUTCHES

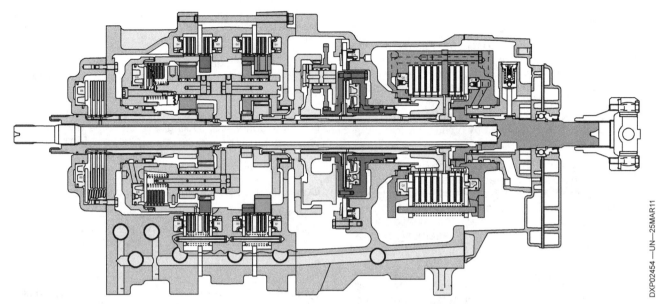

Fig. 23 — 19-Speed Power Shift Transmission

Here is one more example — a 19-speed power shift transmission using planetary gears instead of countershafts (Fig. 23). It contains four clutches and six brakes. We will definitely leave power flows for another time.

A FEW SERVICE NOTES

Hub splines are a potential wear problem in any type of clutch. Worn splines add slack to the system, and it is possible for splines to fail entirely. Worn splines can also make it hard to release a clutch, because the disk can't slide when pressure is removed.

The two principal malfunctions of a clutch are slipping while engaged and dragging while released. Other concerns include vibration, noise, and grabbing. There are too many possible causes to discuss all of them here. Check for incorrect parts, wear, breakage, warping, contamination, linkage defects, misalignment, or incorrect adjustment. Consult the repair manual for the specific machine.

Before disassembly, check the oil level and look for leaking oil.

During repair, check all the oil seals at the clutch release forks, flywheel, and clutch shaft end of the housing. Replace worn seals. If a seal has small cracks on the lips or retainer, it should be replaced. If the seal material appears compressed and is glazed, it is wise to replace it.

The best practice with seals is, "If in doubt, throw it out."

Failure to keep oil in (wet clutch) or out of (dry clutch) the clutch housing will cost a lot more in repairs than the price of an oil seal.

Modern transmissions will need to be recalibrated after any repairs. Minor differences in dimensions can affect clutch piston fill volumes, significantly changing shift quality. Follow calibration instructions.

Here are some things to watch for when servicing clutches:

- Use care when disconnecting the clutch linkage. A bent rod or damaged carrier can result in lost motion, hard operation, or faster wear.
- Don't let the full weight of the transmission rest on the clutch shaft. Use the proper hoists or jacks and keep the parts aligned when separating a machine. Failure to do so will result in damage to the clutch disk and release bearing.
- Mark the pressure plate assembly and flywheel before removing the pressure plate. When assembled, the unit can be matched with the marks and engine balance maintained. When replacing the entire unit, clutch balance is not so critical because the new unit will be static-balanced. Rear crankshaft oil seal failure is often the result of an unbalanced clutch.
- After removing the pressure plate assembly, check these four areas: clutch plate, clutch disk, flywheel plate, and release bearing.

CAUTION: Use proper equipment and methods prescribed by the technical manual when dismantling the pressure plate assembly. Spring pressure must be kept under control and released evenly to avoid personal injury.

DRY CLUTCHES

Always check the clutch release and pressure mechanisms. Pressure should be even all around the clutch. The release levers must be adjusted so that contact with the release bearing will be uniform and the pressure plate will release evenly.

Continued on next page

CLUTCHES

After rebuilding the clutch, check the clutch pedal free travel. When the clutch pedal is released, a retracting spring pulls the pedal back so that the release bearing fork does not contact the release bearing. Another spring pulls the release bearing carrier back so that the bearing does not contact the clutch plate levers. Thus, in normal driving, with the clutch engaged and the driver's foot removed from the clutch pedal, the clutch release bearing is not turning. This is ensured by having the correct free travel on the pedal.

Always check the clutch release setting during clutch service. As the clutch disk facings wear, the clutch disk becomes thinner. The clutch plates then move closer to the flywheel in order to engage the disk. As the pressure plate moves closer to the flywheel, the release fingers move closer to the release bearing. After long operation, these fingers will contact the release bearing and it will run all the time.

Additional wear on the clutch disk cannot be taken up by the pressure plate because the release fingers are contacting the release bearing. The clutch will then start to slip, and the release bearing will wear out rapidly.

Do not use too much lubrication on the clutch pilot bearing, release bearing, or release fork. Excessive lubrication can lead to erratic operation of the clutch.

Some release bearings are sealed units that never need lubrication.

WET CLUTCHES

Wet clutches are designed for extremely long, trouble-free service. They often last the life of the machine with no repairs. Do not disturb a wet clutch assembly unless you are certain something is wrong. Unnecessary tinkering will increase the risk of problems.

Always follow diagnostic procedures step by step before any disassembly. If service is required, follow all instructions to the letter.

Obviously, oil is critical to the life of a wet clutch. Oil provides lubrication, cooling, and often the engagement force. Always use the correct type of oil. Maintain the proper level. Do everything possible to prevent contamination. Replace oil and filters as recommended.

Consider these points when servicing a wet clutch:

- Is the oil being circulated properly?
- On units equipped with pumps, are the lines clogged?
- Is the pump moving its full capacity of oil?
- Is the inlet line or filter dirty?

Clutch disks and other parts may or may not be symmetrical. Always note whether it matters, and make sure you turn the correct side forward. It's frustrating to repeat a major disassembly because a retainer was installed with the snap ring recess on the wrong side or other such errors.

Separator plates may be flat or "wavy." Always use the correct type. When installing wavy plates, follow instructions to make sure they are oriented properly. For example, a tang on each plate may be marked with a notch, and the notches may need to be staggered instead of aligned.

For details, see the machine technical manual.

OTHER TYPES OF CLUTCHES

Besides the familiar disk and plate clutches used in so many different ways, you might encounter a few other types.

CONE CLUTCHES

Cone clutches (Fig. 24) were once common in trucks and cars, but disk and plate clutches perform better. Cones tend to grab excessively. Many synchronizers incorporate cone-type friction elements, although they are not generally considered clutches.

A tapered friction element fits inside a tapered drum. To engage the inner element, it is pushed into the drum, perhaps using a throw-out bearing as shown. The amount of torque transmitted is high relative to the force applied. Smooth engagement is a challenge.

A—Drive Member
B—Driven Member
C—Throw-Out Bearing
D—Driven Shaft
E—Clutch Lining
F—Drive Shaft

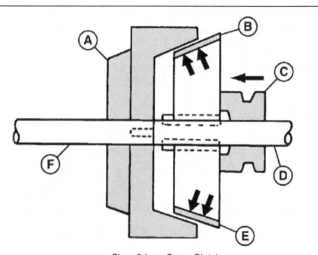

Fig. 24 — Cone Clutch

CLUTCHES

EXPANDING SHOE CLUTCHES

Expanding shoe clutches (Fig. 25) have mostly been replaced by disk and plate clutches. They are reminiscent of drum brakes, with friction "shoes" pressed outward against the drum.

A—Clutch Release Bearing
B—Shoes
C—Pivot Links
D—Outer Member

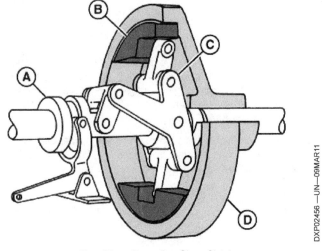

Fig. 25 — Expanding Shoe Clutch

Fig. 26 shows a common mechanism for engaging expanding shoe clutches. Pushing the center tube forces the shoes out to engage the clutch. Pushing the tube "over-center" to the stop will hold it in the engaged position until the control lever is pulled to release it.

A—Released (Locked)
B—On-Center
C—Over-Center (Locked)
D—Stop

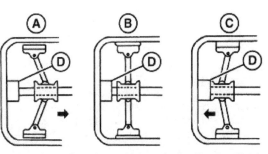

Fig. 26 — Over-Center Linkage for Locking the Clutch While Engaged or Disengaged

Continued on next page

CLUTCHES

CENTRIFUGAL CLUTCHES

The centrifugal clutch (Fig. 27) is a variation of the expanding shoe design. It has flyweights on a rotating shaft. Springs hold the flyweights inward, away from the drum, to release the clutch. As shaft speed increases, centrifugal force overcomes spring force and the shoes push out against the drum. The faster it spins, the more torque it can transmit.

Centrifugal clutches are common in light duty applications from chain saws to recreational vehicles. They are not used in heavy duty equipment for farming and construction.

A—Outer Member
B—Shoe
C—Vane
D—Inner Member

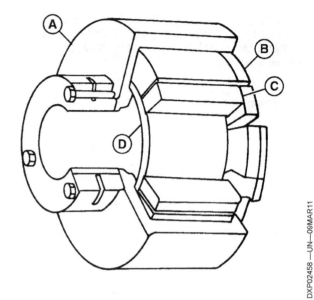

Fig. 27 — Expanding Shoe Clutch (Centrifugal Type)

BAND CLUTCHES

Band clutches are similar to the expanding shoe design, except the friction element is a band wrapped around the outside rather than shoes inside (Fig. 28). This type of device is more commonly used for a brake than for a clutch.

A—Clutch Linkage Arms
B—Band
C—Rotating Member (Flywheel)

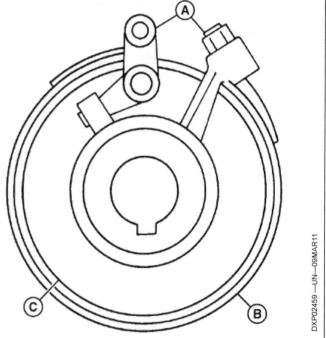

Fig. 28 — Band Clutch

Continued on next page

CLUTCHES

MAGNETIC CLUTCHES

Magnetic clutches (Fig. 29) use electromagnetic force to press the friction surfaces together. A rotating pulley contains the clutch disk. The disk is held a small distance away from the rotor assembly, so the rotor does not turn unless something forces the disk against it.

The clutch is engaged when voltage is supplied to the field assembly. This creates a strong magnetic field that pulls the disk against the rotor. The rotor hub is the output. When engaged, everything except the field assembly is turning.

When the voltage is disconnected, the magnetic field is collapsed. The disk is freed from the rotor and the transmitting of power to the mechanism is instantly stopped.

The most common use for a magnetic clutch is on an air conditioner compressor pulley.

SERVICING MAGNETIC CLUTCHES

If the clutch suddenly loses power and fails to function, look for a failure in the electrical circuit. First check the electrical connections, wires, switch, and the circuit breaker in the switch.

If these components are OK, check the voltage to the field and amperage of the field, using a battery eliminator or a storage battery and volt/ohmmeter.

Check the specifications on each clutch for correct input, amperage, and resistance according to the machine technical manual.

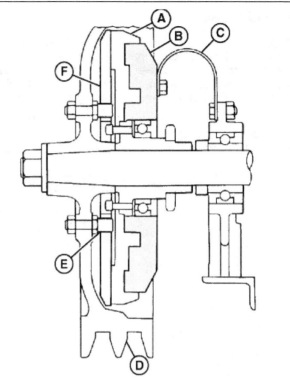

Fig. 29 — Magnetic Clutch

A—Rotor Assembly
B—Field Assembly
C—Connecting Strap
D—Rotating Pulley
E—Drive Studs
F—Clutch Disk

Continued on next page

CLUTCHES

SLIP CLUTCHES

A slip clutch is a device to protect against excessive torque, which could cause machine damage (Fig. 30). It has the same function as a shear pin or an electrical fuse. If a machine is overloaded, the slip clutch should yield before machine components are damaged.

A slip clutch must be properly adjusted. If it slips too easily, it reduces machine capability and may quickly wear out. If it doesn't slip when it should, it isn't serving its purpose.

A—Adjusting Spring
B—Clutch Facing
C—Revolving Plate
D—Two-Piece Drive Shaft

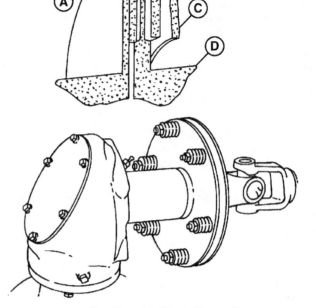

Fig. 30 — Slip Clutch (Engaged)

Continued on next page

CLUTCHES

OVERRUNNING CLUTCHES

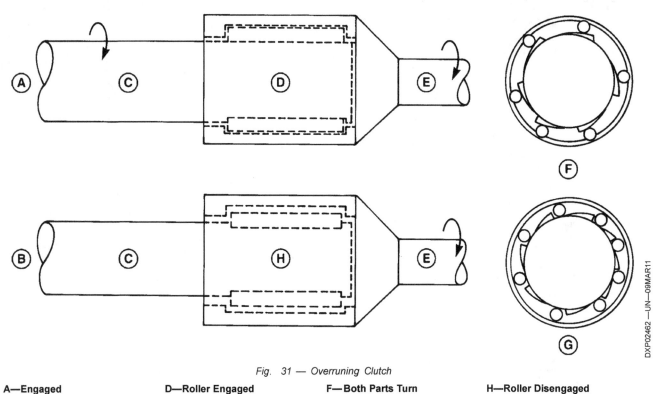

Fig. 31 — Overrunning Clutch

A—Engaged
B—Disengaged
C—Driveline
D—Roller Engaged
E—Driven Line
F—Both Parts Turn
G—Driveline Idles
H—Roller Disengaged

An overrunning clutch is essentially a ratchet. It does not engage and disengage; it merely allows the output shaft to continue to rotate if the input shaft stops. The most common use is in the driveline for implements with high inertia, such as a rotary mower or hay balers. A heavy flywheel needs time to slow down.

Fig. 31 shows a common type of overrunning clutch. It utilizes rollers trapped between a smooth bore on one side, and small ramps on the other. When the input shaft is rotating, the rollers are wedged tightly between the ramps and the bore. This forces the output shaft to rotate.

When the input shaft stops, momentum will cause the output shaft to continue to rotate. The rollers move down their ramps and no longer lock the two shafts together. The load can then coast to a stop.

There are other types of overrunning clutches that use springs, sprags, or cams. The operation of these other types is similar.

It is possible but not common to use an overrunning clutch in a two-speed overdrive. This application allows the engagement of a higher gear without disengaging a lower one. The lower gear merely free-wheels until the higher gear is disengaged; the lower gear then resumes carrying the load.

In the past, certain cars had overrunning clutches, which could be locked to provide engine braking or released for coasting.

PNEUMATIC (AIR) CLUTCHES

The air-actuated clutch is a clutch pack that contains a small rubber tube. This tube is positioned behind a driven member and a fixed member.

To engage the clutch, a valve is opened and air is forced into the tube, causing it to expand. This expansion forces the driven plate against the drive plate. The higher the air pressure, the greater the force exerted by the tube against the driven plate. Air pressure is decreased to disengage the clutch.

This clutch provides smooth engagement because the rubber tube absorbs the shocks.

HYDRAULIC CLUTCHES

Hydraulic clutches work in much the same way as the pneumatic clutch, but use oil instead of air.

Oil is forced into a chamber between the fixed and moving driven members. A valve is then opened, allowing oil to flow into the chamber. As oil pressure builds up, the walls of the chamber are forced outward, causing contact between the driven and drive members. The amount of pressure exerted on the clutch disk is regulated by the pressure of the oil in the chamber.

Continued on next page

CLUTCHES

These clutches are also fairly shock resistant.

CLUTCHES

TROUBLESHOOTING

The eight basic clutch troubles are:

- Chattering: especially in low or reverse speeds.
- Dragging: or failure to release promptly and fully, thus making gears hard to shift.
- Squeaks: particularly when pedal is depressed.
- Rattles: especially at low speeds or standing.
- Grabbing: violent and sudden engagement.
- Slipping: failure to transmit full power.
- Vibrations at either high or low speed or periodically.
- Failure to transmit power at all.

Note that different clutch problems can have the same cause. For example, oil or grease on the clutch disk facings can cause the clutch to slip, but may also make it chatter, drag, or grab.

CHATTERING

1. Oil or grease on clutch disk facings.
2. Glazed or worn facings.
3. Worn, loose, or spongy engine mountings.
4. Worn or loose splines in clutch hub or on clutch shaft.
5. Wear or looseness in universal joint, differential, or drive axles.
6. Splined disk hub sticking on splined shaft.
7. Cracked or scored pressure plate or flywheel face.
8. Warped clutch disk.
9. Warped pressure plate.
10. Pressure plate sticking on driving studs.
11. Sticking or binding release levers.
12. Unequally adjusted release levers.
13. Unequal length or strength of clutch springs.
14. Bent clutch shaft.
15. Misalignment of power train.

DRAGGING

1. Oil or grease in clutch.
2. Warped clutch disk.
3. Splined disk hub sticking on splined shaft.
4. Sticking pilot bearing or bushing.
5. Sticking release sleeve.
6. Warped pressure plate or clutch cover.
7. Broken disk facing.
8. Accumulation of dust in clutch.
9. Incorrect clutch or pedal adjustment.

10. Engine idling too fast.
11. Misalignment of parts.

SQUEAKS

1. Clutch release bearing needs lubrication.
2. Pilot bearing in flywheel needs lubrication.
3. Release sleeve needs lubrication.
4. Misalignment.

RATTLES

1. Loose hub in clutch disk.
2. Worn release bearing.
3. Worn release part.
4. Worn pilot bearing.
5. Worn splines in hub or on shaft.
6. Worn driving pins in pressure plate.
7. Wear in transmission or driveline.
8. Worn transmission bearings.
9. Bent clutch shaft.
10. Unequal adjustment of release levers.
11. Misalignment.

GRABBING

1. Oil or grease on clutch disk facings.
2. Splined hub sticking on splined shaft.
3. Pressure plate sticking on drive studs.
4. Glazed or worn facings.
5. Sticking or binding release levers.
6. Sticking or binding clutch pedal or linkage.
7. Misalignment.

SLIPPING

1. Worn clutch disk facings.
2. Weak or broken springs.
3. Improper clutch or pedal adjustment.
4. Oil or grease on facings.
5. Warped disk.
6. Warped pressure plate.
7. Sticking release levers.
8. Pressure plate sticking or binding on studs.
9. Misalignment.

Continued on next page

CLUTCHES

VIBRATIONS

1. Bent clutch shaft.
2. Defective clutch disk.
3. Dust in clutch.
4. Improper assembly of clutch to flywheel.
5. Use of rigid disk instead of flexible type.
6. Unmatched springs in pressure plate.
7. Misalignment of parts.

FAILURE

1. Clutch disk hub torn out.
2. Friction facings on clutch disk torn off or worn out.
3. Broken springs.
4. Incorrect adjustment of pressure plate.
5. Improper adjustment of clutch or pedal.
6. Splined disk hub stuck on splined shaft.

SUMMARY: TROUBLESHOOTING OF CLUTCHES

You can locate the basic problems using the charts here. Refer to your technical manual for specific problems that you might face in servicing a clutch.

TEST YOURSELF

QUESTIONS

1. Name four types of clutches.
2. What are the two kinds of disk clutches?
3. What characteristics must a clutch facing have? Name two.
4. How can you tell a rigid clutch disk from a flexible disk?
5. (Fill in the blanks.) An overrunning clutch will _____ in one direction, but will _____ in the other.
6. (True or False?) In an over-center linkage, it is necessary to hold your foot on the pedal to keep the clutch disengaged.
7. (True or False?) The clutch release bearing should rotate all the time in a plate clutch.
8. Grease or oil on a dry clutch disk surface usually causes:
   a. Grabbing.
   b. Slipping.
   c. Both of the above.

MANUAL TRANSMISSIONS

INTRODUCTION

A transmission is a system of gears and shafts used to transmit engine power to the drive axle. A manual transmission is one in which the operator manually shifts the speeds, as opposed to power shift transmissions, which will be discussed in chapter 4.

The transmission has two primary functions:

1. Selects the direction of travel.

2. Selects a gear ratio to provide the appropriate travel speed.

The transmission can also power special drive adapters for PTO, MFWD, hydraulic pump, etc.

The transmission is usually located behind the engine and in front of the differential.

MANUAL TRANSMISSIONS

TYPES OF SHIFTERS

Three types of mechanisms are used to engage gears:

- Sliding gear
- Collar shift
- Synchronizer

It isn't unusual for two different types of shifters, or even all three, to be used in the same transmission.

SLIDING GEAR

As the name implies, this type of shifter operates by sliding a gear on a shaft. The gear is splined to the shaft, so they rotate together.

However, the gear can be slid along the splines on the shaft so that the gear teeth do or do not engage the teeth on another gear on another shaft.

Sliding gear transmissions are the least expensive type. They are used primarily on low-speed and/or light-load applications. They are not suitable for shifting on-the-go, and gear teeth can be damaged by improper shifting.

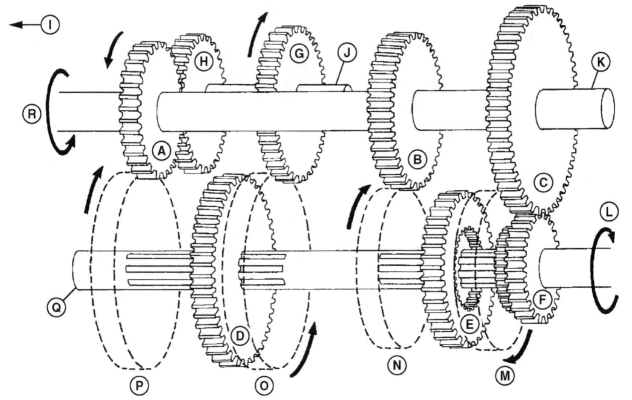

Fig. 1 — Sliding Gear Transmission with Parallel Shafts

A—Gear A	G—Gear G	L—Power to Differential	P—1st Gear
B—Gear B	H—Gear H	M—3rd Gear	Q—Output Shaft
C—Gear C	I—Forward	N—2nd Gear	R—Power from Engine
D—Gear D	J—Reverse Idler Shaft	O—Reverse Gear	
E—Gear E	K—Input Shaft		
F—Gear F			

Fig. 1 illustrates a sliding gear transmission that provides three forward speeds and one reverse. Let's look closely at the design.

The input shaft provides power from the engine. If the clutch is engaged, it runs all the time. Gears A, B, and C are mounted rigidly on the input shaft. They run at all times.

The reverse idler shaft is used only when the transmission is in reverse. Gears G and H are mounted rigidly on the reverse idler shaft. Gear H is in mesh with Gear A, so gears G and H turn all the time, but in the opposite direction from the input shaft.

The output shaft provides power to the differential. The direction and speed at which it turns will depend on which gears are engaged. Three gears are mounted on the output shaft:

- Gear D is a sliding gear. It is splined to the output shaft. It can be slid forward to engage Gear A or rearward to engage Gear G.
- Gear E is another sliding gear. It is also splined to the output shaft. It can be slid forward to engage Gear B. But the hub of Gear E is unique. The rear half has larger splines to fit over the front of Gear F.

Continued on next page

- Gear F is unlike any of the others. It rotates freely, or "floats," on the output shaft. Gear F is always in mesh with Gear C, so it turns all the time.

If neither Gear D nor Gear E is engaged with another gear, the transmission is in neutral. The input shaft turns, but the output shaft does not.

For 1st speed, we slide Gear D forward to engage Gear A. Gear A turns all the time, so this turns Gear D, which is splined to the output shaft. A small gear is driving a large gear, so this is our slowest speed.

For reverse, we slide Gear D rearward to engage Gear G. Remember, Gear G turns all the time, but in the opposite direction from the input shaft. This turns Gear D and the output shaft, but in the opposite direction from before.

For 2nd speed, we slide Gear E forward to engage Gear B. This is exactly like 1st, except now a medium-size gear is driving a medium-size gear. This produces an intermediate speed.

For 3rd speed, we slide Gear E rearward to fit over the splines on the front of Gear F. Remember, Gear F turns all the time, but it floats on the output shaft. By engaging splines on Gear F and splines on the output shaft, Gear E locks them together. A large gear (C) is driving a small gear (F), so this is our fastest speed.

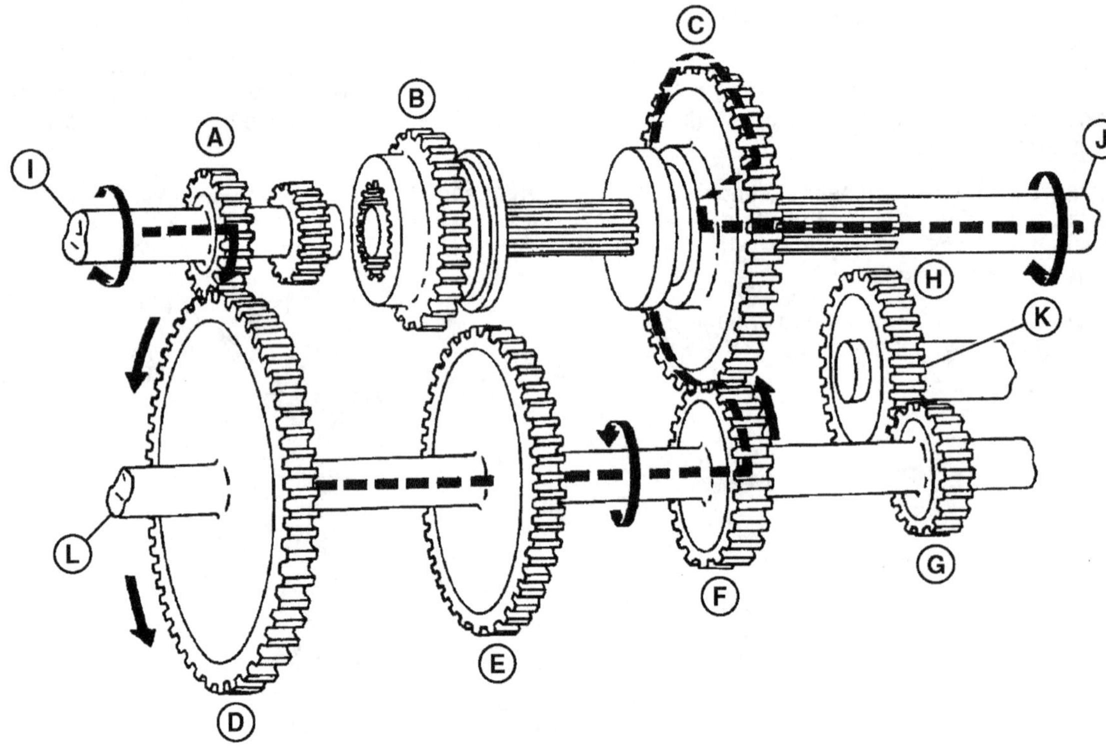

MANUAL TRANSMISSIONS

Fig. 2 — Sliding Gear Transmission with Shafts in Line (1st Gear Shown)

A—Gear A
B—Gear B
C—Gear C
D—Gear D
E—Gear E
F—Gear F
G—Gear G
H—Gear H
I—Input Shaft
J—Output Shaft
K—Reverse Idler Gear
L—Countershaft

Fig. 2 illustrates a different type of sliding gear transmission. In this one, the output shaft is directly in line with the input shaft. Power flows through a countershaft, then through a second gear mesh to a gear on the output shaft.

The transmission is shown in 1st speed. The input shaft is driven by the engine, turning Gear A. Gear A is always in mesh with Gear D. Gears D, E, F, and G are all mounted rigidly on the countershaft, so they turn all the time. In 1st speed, we slide Gear C forward to engage Gear F. This turns the output shaft. A small gear (F) is driving a large gear (C), so this is our slowest speed.

For reverse, we slide Gear C rearward to engage Gear H. (Although it doesn't appear from the perspective of this diagram that the gear teeth would align, they do.) Gear H is the reverse idler. It is always in mesh with Gear G, so it turns all the time, but in the opposite direction from the countershaft.

For 2nd speed, we slide Gear B rearward to engage Gear E on the countershaft. This is exactly like 1st, except now a medium-size gear is driving a medium-size gear. This produces an intermediate speed.

For 3rd speed, we slide Gear B forward. Similar to the previous example, the hub of Gear B has a unique design. The rear half is splined to the output shaft. The front half has larger splines to fit over the back of Gear A. This locks the output shaft to the input shaft for a straight-through drive. This is a common design in automotive transmissions. Because power is not being transmitted through any gear meshes, it reduces friction loss and noise. There is no side load on the bearings. This is a very efficient arrangement.

Unlike the previous illustration, Fig. 2 shows how the sliding gears are moved. Shifter forks would fit into the slots on Gears B and C. The shifter forks slide the gears back and forth.

Continued on next page

COLLAR SHIFT

In a collar shift transmission, the gears are in mesh at all times. This reduces the chance of damaging gear teeth. This is a stronger design, especially for high speed and/or heavy load.

The gear itself (Fig. 3) is not splined to the shaft. It rotates freely, or floats, on the shaft. Right beside the gear is a sliding collar which is splined to the shaft by a shifter gear. When we slide the collar forward or back to lock onto a gear, this locks the gear to the shaft. This is exactly the same principle we used for 3rd speed in both of the previous examples.

As before, shifter forks fit into grooves on the collars to slide them back and forth.

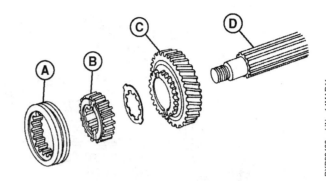

Fig. 3 — Typical Shifter Collar and Mating Gear

A—Shifter Collar
B—Shifter Gear (Splined to Shaft)
C—Driven Gear (Rides Free on Shaft)
D—Shaft

Continued on next page

MANUAL TRANSMISSIONS

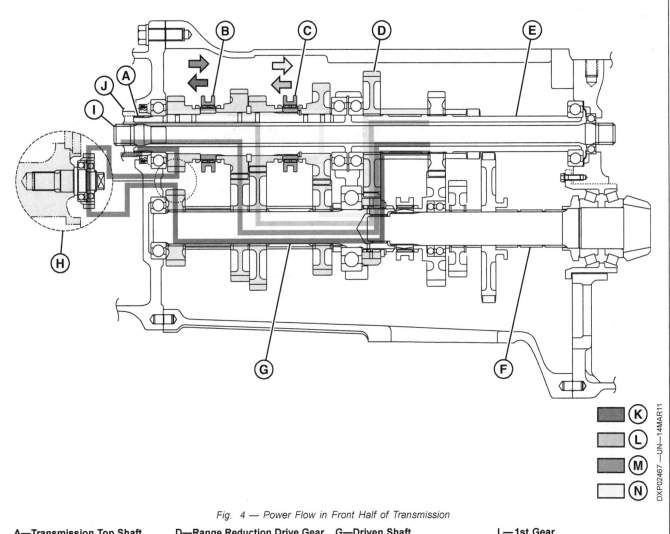

Fig. 4 — Power Flow in Front Half of Transmission

A—Transmission Top Shaft
B—Reverse and 2nd Gear Shift Collar
C—1st and 3rd Gear Shift Collar
D—Range Reduction Drive Gear
E—Range Reduction Shaft
F—Differential Drive Shaft
G—Driven Shaft
H—Reverse Idler
I—PTO Shaft
J—Traction Clutch Shaft
K—Reverse Gear
L—1st Gear
M—2nd Gear
N—3rd Gear

The transmission (Fig. 4) incorporates five separate shafts. No two are connected except through gears, and all five can be turning at different speeds. The PTO shaft is only passing through the hollow top shafts; it has no involvement in transmission operation.

The front half of the transmission provides three forward speeds and one reverse. We'll discuss the rear half in a moment.

Unless the operator is holding the clutch pedal down, the traction clutch shaft is turning at engine speed. If both shift collars (forward and reverse) are in neutral, nothing else turns.

Either shift collar can be slid in either direction to lock a gear to the traction clutch shaft. The different gear ratios drive the countershaft at different speeds. Because the front gear set includes a reverse idler, it turns the countershaft in the opposite direction.

All five gears on the countershaft are splined to the shaft, including one at the very back that drives the range reduction drive gear, which is splined to the range reduction shaft. Therefore, we can turn the range reduction shaft in any of three speeds forward or one speed reverse.

Continued on next page

MANUAL TRANSMISSIONS

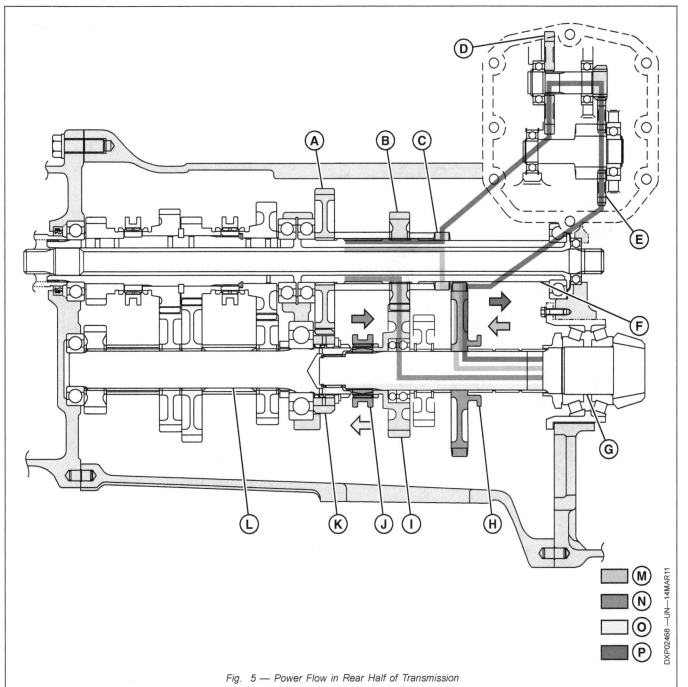

Fig. 5 — Power Flow in Rear Half of Transmission

A—Range Reduction Drive Gear
B—B Range Drive Gear
C—A Range Drive Gear
D—Creeper Drive Gear
E—Creeper Driven Gear
F—Range Reduction Shaft
G—Differential Drive Shaft
H—A Range and Creeper Shift Sliding Gear
I—B Range Driven Gear
J—B and C Range Shift Collar
K—C Range Driven Gear
L—Driven Shaft
M—A Range
N—B Range
O—C Range
P—Creeper Range

The rear half of the transmission (Fig. 5) provides three ranges. If equipped with the optional creeper, there are four ranges. All the previous choices are available in each range, so the transmission provides nine speeds forward and three reverse. If equipped with creeper, make that twelve speeds forward and four reverse.

The A Range is engaged by a sliding gear instead of a collar shift. This is suitable for the low speeds involved.

When we slide the gear forward, it engages the small gear on the range reduction shaft. This turns the differential drive shaft at slow speed.

If equipped with creeper, we use the same sliding gear. Sliding it rearward engages the creeper-driven gear. This turns the differential drive shaft at extremely slow speed.

Continued on next page

OUO1082,000136A -19-16MAY11-6/18

3-7

MANUAL TRANSMISSIONS

B and C Ranges are engaged by a collar shift. B Range driven gear floats on the differential drive shaft. It is in mesh with the large gear on the range reduction shaft. Sliding the shift collar rearward locks the gear onto the shaft for medium speed. Sliding the shift collar forward locks the differential drive shaft to the shaft in front of it for fastest speed.

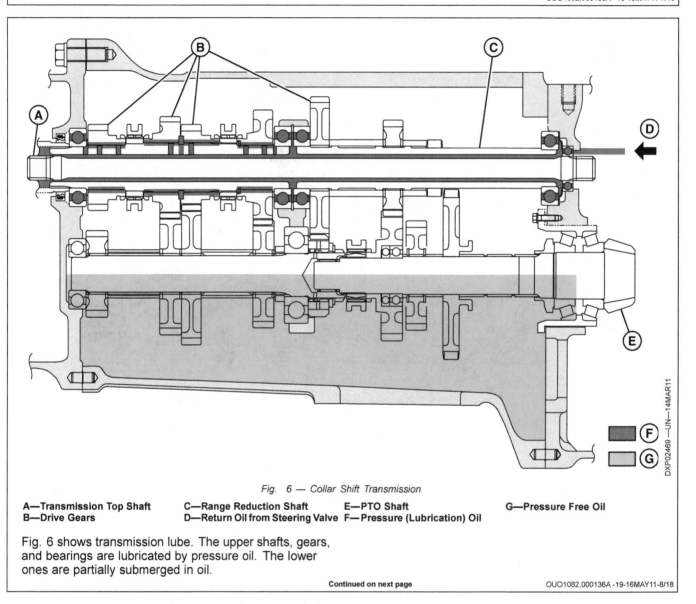

Fig. 6 — Collar Shift Transmission

A—Transmission Top Shaft
B—Drive Gears
C—Range Reduction Shaft
D—Return Oil from Steering Valve
E—PTO Shaft
F—Pressure (Lubrication) Oil
G—Pressure Free Oil

Fig. 6 shows transmission lube. The upper shafts, gears, and bearings are lubricated by pressure oil. The lower ones are partially submerged in oil.

Continued on next page

MANUAL TRANSMISSIONS

SYNCHRONIZER

A synchronizer is a collar shift with one added feature. It matches the speeds of the collar and the gear before locking them together, to prevent "clash." This allows shifting on-the-go. A synchronizer is helpful even when the machine is stopped, because shafts can be spinning in neutral.

Synchronizers come in various designs. All of them perform the same function. Examples of synchronizers are cone-type synchronizer, pin synchronizer, and disk and plate synchronizer.

Fig. 7 is a simplified illustration of a cone-type synchronizer. The hub and sleeve are splined to the shaft. The blocking rings rotate with the assembly, but can be deflected slightly in either direction.

The inside of each blocking ring is a tapered friction surface. It fits over a matching surface on the gear. In neutral, these surfaces do not touch. When pushed together, the surfaces function like a cone clutch mentioned in chapter 2.

Let's walk through an example. Assume the machine is stopped. The shaft is spinning, but the gears are stopped. Now the operator pushes the clutch pedal and attempts to shift into gear. Though the shaft is no longer driven by the engine, it continues to spin due to momentum.

As soon as the shift lever pushes the sleeve, the blocking ring contacts the gear. Friction acts as a brake on the blocking ring, turning it back slightly against spring force. This misaligns the splines, and the sleeve cannot move any farther until the blocking ring is centered again. That's the synchronizer's first job, to prevent clash engagement.

Friction between the blocking ring and the gear then acts as a clutch to equalize speeds. It will either stop the shaft or accelerate the gear until both are turning the same speed. That's the synchronizer's second job.

Once the speeds are equalized, the gear will no longer be holding back on the blocking ring. Spring force will center it, so the splines are aligned again. Now the sleeve can be slid over the splines on the blocking ring and gear to engage the gear without clashing. That's the synchronizer's third and most important job.

A—Synchronizer in Neutral Position Before Shift
B—During Synchronization—Blocking Ring and Gear Shoulder Contacting
C—Shift Completed—Collar Locks Driven Gear to Hub and Shaft
D—Blocking Ring
E—Synchronizer Sleeve
F—Driven Gear
G—Hub
H—Contact Here
I—Meshing Occurs Here
J—Power Flow

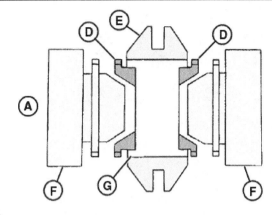
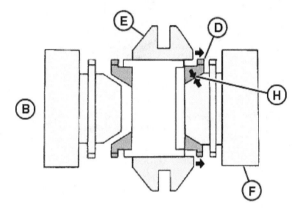
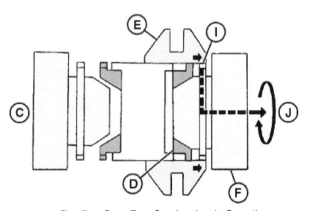

Fig. 7 — Cone-Type Synchronizer in Operation

MANUAL TRANSMISSIONS

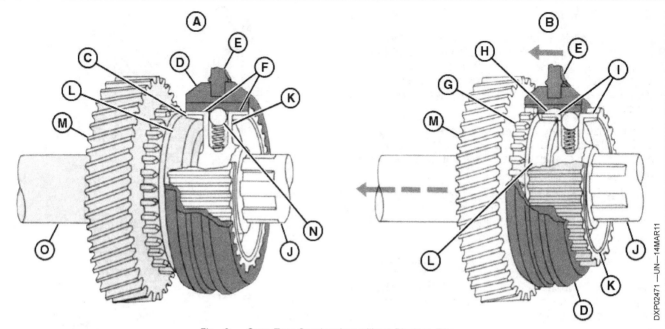

Fig. 8 — Cone-Type Synchronizer without Blocking Rings

A—Moving Toward Engagement
B—Fully Engaged
C—External and Internal Cones Contacting
D—Sliding Sleeve
E—Shift Fork
F—Internal Cones
G—External and Internal Teeth Engaged
H—Spring-Loaded Ball Out of Groove
I—Internal Cones
J—Input Shaft
K—Hub
L—External Cone
M—Driven Gear
N—Spring-Loaded Ball in Groove in Sliding Sleeve
O—Output Shaft

There are variations on the cone synchronizer. Fig. 8 shows a light-duty design with no blocking rings. The hub itself provides a friction surface for the cone clutch. The hub can slide a short distance in either direction on the shaft splines. Note the spring-loaded detent ball in the hub, seated in a groove inside the sleeve. Before you can slide the sleeve over to engage the gear, you must exert enough force to unseat the detent ball. It is assumed this will push the cones together enough to equalize speed. This design would be less expensive than the previous example, but not as durable or as positive.

Continued on next page

MANUAL TRANSMISSIONS

Fig. 9 shows another variation that uses pins for the blocker function. The outer cones fit over splines on the sides of the gears. The sleeve fits over splines on a center hub, which is not shown. To engage a gear, you push the sleeve out over the same splines the outer cone is on, locking the gear to the shaft.

But first, the sleeve will push the friction ring against the outer cone. In order to engage a gear, the tapered shoulders on the friction ring pins must slide into the chamfered holes in the sleeve. If the cone and sleeve are not turning at the same speed, friction will hold the small ends of the pins against the sides of the holes and prevent engagement. This performs exactly the same job as a blocker ring. As soon as speeds equalize, the pins slide in easily and you can engage the gear.

This type of synchronizer is used for 1st and 3rd gears in the transmission shown in Fig. 11.

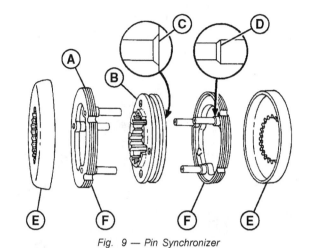

Fig. 9 — Pin Synchronizer

A—Grooves on Friction Surface
B—Sleeve
C—Sleeve Hole Chamfer
D—Pin Chamfer
E—Outer Cone
F—Friction Ring and Pin Assemblies

Continued on next page

MANUAL TRANSMISSIONS

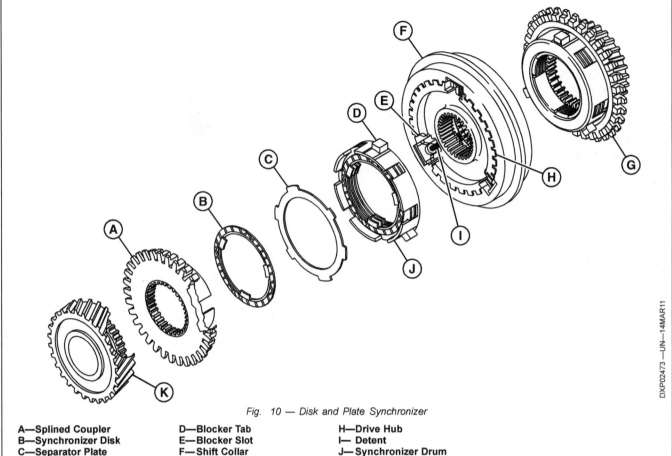

Fig. 10 — Disk and Plate Synchronizer

A—Splined Coupler
B—Synchronizer Disk
C—Separator Plate
D—Blocker Tab
E—Blocker Slot
F—Shift Collar
G—2nd Speed Drive Gear
H—Drive Hub
I—Detent
J—Synchronizer Drum
K—Reverse Drive Gear

Fig. 10 shows a disk and plate synchronizer. It performs the same functions of blocking, speed matching, and engaging. Instead of a cone clutch for speed matching, it uses multiple disks and plates. This design is suitable for heavier service. It's the type used in most large transmissions.

A disk and plate synchronizer is used for 2nd and reverse gears in the example transmission in Fig. 11.

Continued on next page

MANUAL TRANSMISSIONS

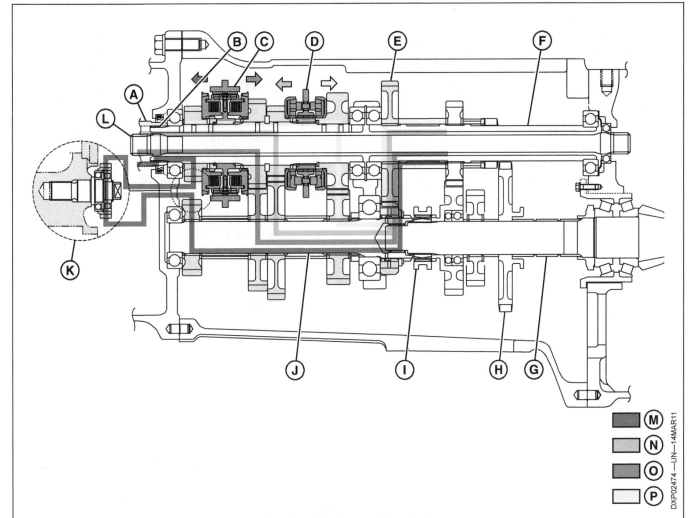

Fig. 11 — Synchronized Transmission

A—Traction Clutch Shaft
B—Transmission Top Shaft
C—Reverse and 2nd Gear Synchronizer
D—1st and 3rd Gear Shift Synchronizer
E—Range Reduction Drive Gear
F—Range Reduction Shaft
G—Differential Drive Shaft
H—A Range and Creeper Shift Sliding Gear
I—B and C Range Shift Collar
J—Driven Shaft
K—Reverse Idler
L—PTO Shaft
M—Reverse Gear
N—1st Gear
O—2nd Gear
P—3rd Gear

This synchronized transmission (Fig. 11) is identical to the one previously described, except two shift collars are replaced by synchronizers. Also note that all three types of shifters are used in one transmission — synchronizers for speeds, collar shift for B and C ranges, sliding gear for A range. We can shift speeds on-the-go, but must stop to shift ranges.

A synchronizer must:

- Block engagement if speeds do not match.
- Accelerate or decelerate one component to achieve equal speeds.
- Allow engagement once the speeds match.

Continued on next page

MANUAL TRANSMISSIONS

There are endless variations in transmission designs. Fig. 12 shows a 3-speed design that uses a sliding gear for 1st and reverse plus a synchronizer for 2nd and 3rd. For 3rd speed, it locks the input and output shafts together.

A—Constant Mesh Gear
B—Synchronizer
C—Sliding Gear
D—Output Shaft
E—Countershaft
F—Input Shaft

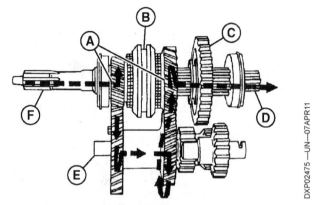

Fig. 12 — 3-Speed Synchromesh Transmission (Power Flow for 2nd Gear Shown)

Continued on next page

MANUAL TRANSMISSIONS

Fig. 13 — 8-Speed, Partially Synchromeshed Transmission in a Wheel Tractor

Fig. 13 shows an 8-speed transmission using an offset countershaft on the far side. This basic design was in production for more than 35 years. The bottom shaft uses collar shifters to provide four ranges. The top shaft uses synchronizers to provide two speeds forward and one reverse within each range.

(Restrictions built into the shift linkage prevent engaging the higher reverse speeds, but the transmission is capable of producing them.)

Continued on next page

MANUAL TRANSMISSIONS

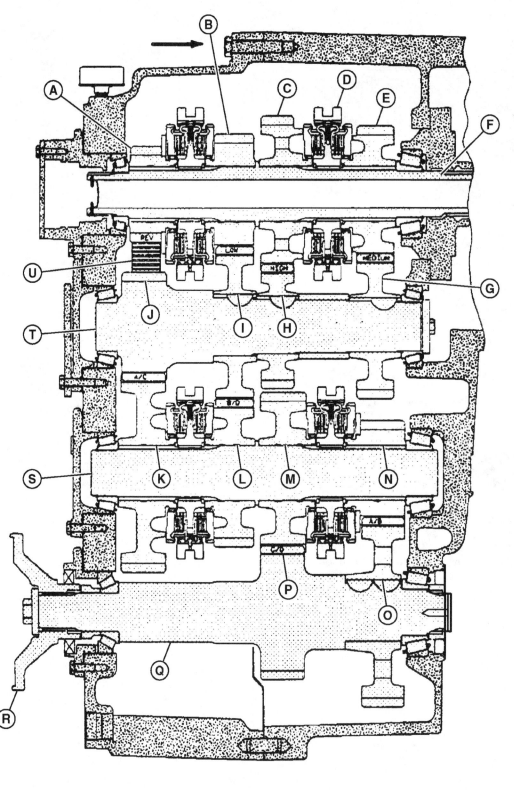

Fig. 14 — 12-Speed Transmission for a Very Large Tractor

MANUAL TRANSMISSIONS

A—Reverse Pinion
B—Low-Speed Pinion
C—High-Speed Pinion
D—Synchronizer
E—Medium-Speed Pinion
F—Top Shaft
G—Medium Speed Gear
H—High-Speed Gear
I—Low-Speed Gear
J—Reverse Speed Gear
K—A and C Range Gear Pinion
L—B and D Range Gear Pinion
M—C and D Range Gear Pinion
N—A and B Range Gear Pinion
O—A and B Range Output Gear
P—C and D Range Output Gear
Q—Transmission Output Shaft
R—Rear Output Yoke
S—Lower Countershaft
T—Upper Countershaft
U—Reverse Idler

Fig. 14 shows a 12-speed transmission for very large four-wheel-drive tractors. It uses four shafts plus a reverse idler. All shifts are synchronized. The lower two shafts provide four ranges. The upper two shafts provide three forward speeds and one reverse in each range.

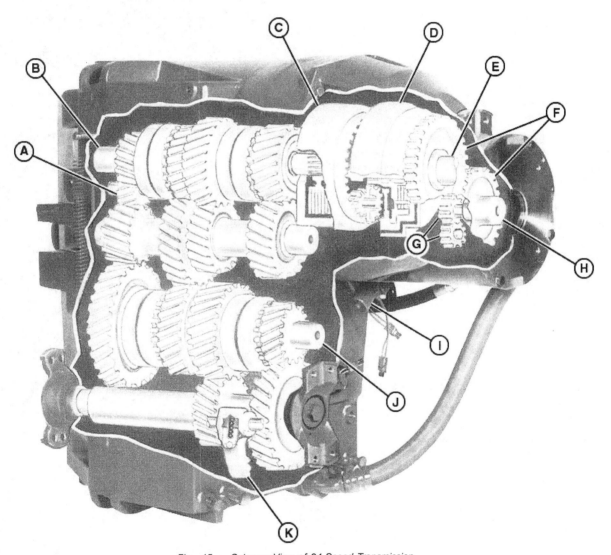

Fig. 15 — Cutaway View of 24-Speed Transmission

A—Reverse Idler Gear
B—Transmission Top Shaft
C—Traction Clutch
D—Two-Speed Planetary
E—Transmission Input Shaft
F—Oil Pump Gear Set
G—Oil Pump Gears
H—Gear Hub
I—Two-Speed Solenoid
J—Lower Countershaft
K—Park Pawl

In the machine (Fig. 15), this becomes a 24-speed transmission. A planetary hi-lo and the traction clutch are installed on the input shaft. Note the large vertical drop between input and output shafts. This is the difference between engine height and axle height.

MANUAL TRANSMISSIONS

SHIFT CONTROLS

Something has to engage these sliding gears and shifter collars and synchronizers. Generally, there is a shifter fork in a slot on the movable component (Fig. 16). Each shifter fork is generally attached to a shifter fork shaft, often called a shift rail. We move the shift rails forward or back to change gears.

A—Shifter Fork
B—Shifter Fork Shaft
C—Collar Gear
D—Shifter Collars

Fig. 16 — Shifter Forks In Collar Shift Transmission

The simplest way to move shift rails is with the tip of the shift lever (Fig. 17). The lever pivots near the lower end, and the tip engages a shift rail on either side.

A—Gear Shift Levers
B—Gear Shift Rails
C—Gear Shift Lever Ball (Pivot Point)

Fig. 17 — Direct Transmission Shifting Mechanism

Continued on next page

MANUAL TRANSMISSIONS

Fig. 18 shows a typical arrangement for an automobile or small tractor. The tip of the shift lever can engage any of three shift rails and slide it back and forth.

A—Reverse Rail
B—Third and Fourth Rail
C—Low and Second Rail
D—Slot for Gearshift Lever
E—Shifter Rails
F—Shifting Pattern

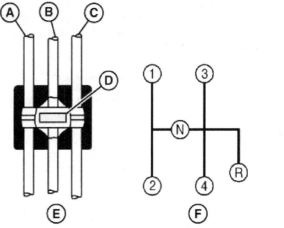

Fig. 18 — Shift Pattern on a 4-Speed Transmission Shown

We need to prevent unintended movement of shift rails. The transmission must not jump out of gear or drift into partial engagement. We use detents, identified as "rail-locking ball and spring" in Fig. 19, to provide positive positioning. Each shift rail has three detented positions. The detent ball clicks into a notch for each position. It will hold the shift rail in that position until you apply enough force to overcome the detent and move the shift rail to one of the other positions.

Also note the two "interlock balls" between each pair of shift rails. They prevent engaging two gears at the same time. Each shift rail has a notch in the neutral position. If either lever is out of neutral, a ball is pushed into the notch on the other lever, locking it in neutral.

This particular transmission has an additional interlock feature which is quite unusual. Note the interlock pin and five balls near the top of the illustration. If the right-hand lever is in neutral, you can't put the left-hand lever in park. If the left-hand lever is in park, you can't move the right-hand lever.

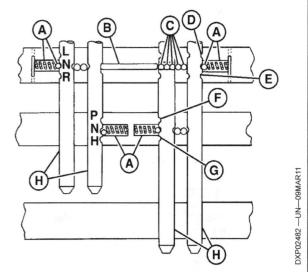

Fig. 19 — Locking System for Shift Rails

A—Rail-Locking Ball and Spring
B—Interlock Pin
C—Interlock Balls
D—1–5
E—2–6
F—4–8
G—3–7
H—Shift Rails

Continued on next page

MANUAL TRANSMISSIONS

On larger machines, it is not practical to have shift levers mounted on top of the transmission. We can operate the shift rails with pivot arms controlled by cables connected to shift levers in the cab (Fig. 20).

A—Shifter Lever
B—Cable
C—Shifter Shaft
D—Shifter Cam
E—Shift Collar
F—Shifter Fork

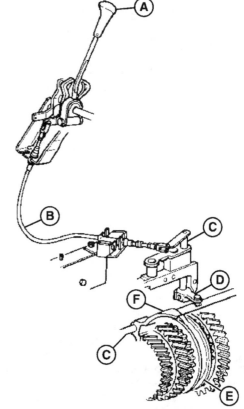

Fig. 20 — Shift Cable for Remote Lever

CAM SHIFTERS

Large transmissions sometimes use a shifter cam to control two or more shift rails (Fig. 21). Rollers on the shift rails follow slotted paths in the cam as it pivots. The cam is operated by a cable to a shift lever in the cab.

A—Shifter Cam
B—Shifter Rollers
C—Shifter Forks
D—Shifter Rail
E—Cam Pivot Point

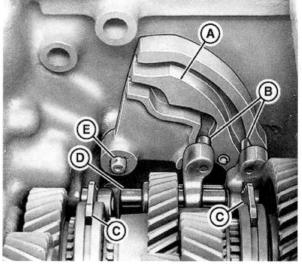

Fig. 21 — Shifter Cam and Rail Assembly

Continued on next page

MANUAL TRANSMISSIONS

One benefit of a shifter cam is straight-line shifting. Fig. 22 shows the shift levers for the 24-speed transmission. Each lever has five positions in a line. No "H" pattern is needed. There is a detent notch on the cam for each position.

A—Speed
B—Range
C—Hi-Lo

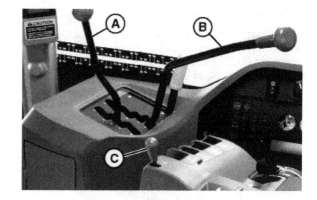

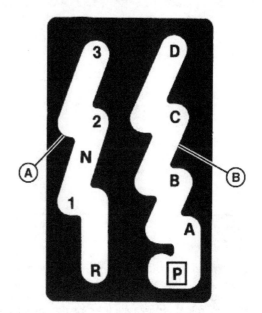

Fig. 22 — Straight-Line Shift Pattern

PARK LOCK

A manual transmission may or may not include a park mechanism. Some rely on a handbrake instead. Some provide a park function by engaging two gears at the same time; the different ratios prevent anything from turning.

Fig. 23 shows one type of park mechanism. A park pawl attached to the transmission housing pivots to engage teeth on a gear on the output shaft. If the shaft can't turn, the machine can't move.

A—Shifter Cam
B—Output Gear
C—Park Pawl
D—Spring
E—Pin
F—Roller

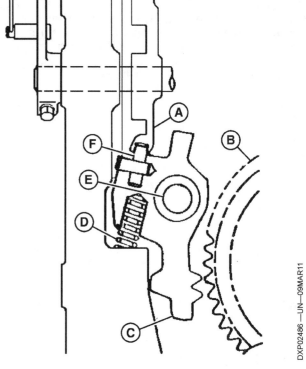

Fig. 23 — Park Pawl

ADJUSTMENTS

Proper assembly of a transmission involves numerous adjustments. Shaft bearings require the correct end play or preload, as discussed in Chapter 1. Shift linkage adjustment is critical, especially the position of shifter forks on the shift rails.

All of this is highly machine-specific. Always follow repair instructions for the machine you are servicing, rather than broad guidelines for transmissions in general. The same is true for maintenance and troubleshooting.

GENERAL MAINTENANCE

Manual transmissions need little maintenance other than periodic oil changes. However, when you repair a transmission, examine the whole gear train to locate worn or faulty parts and repair or replace them at that time. Thus, you may prevent a breakdown and the need to disassemble the transmission once again.

Some things to LOOK for:

- Excessive gear tooth wear or broken teeth.
- Worn-out bearings.
- Broken or distorted shifter detent springs or scored or flat detent balls.
- Damaged or plugged transmission oil lines or passages in shafts.
- In sliding gear transmissions, badly worn splines.
- In collar shift transmissions, excessively worn teeth in shifter collars and hubs.
- In synchromesh transmissions, scored contact surfaces on blockers, drums, or disks.

MANUAL TRANSMISSIONS

TROUBLESHOOTING

This troubleshooting chart is given as a general guide to common transmission failures. The chart lists what the cause might be and the remedy.

MANUAL TRANSMISSION TROUBLESHOOTING	
POSSIBLE CAUSE	REMEDY
TRANSMISSION NOISY IN NEUTRAL	
Transmission not aligned with engine	Align
Bearings dry, badly worn, or broken	Lubricate or replace
Transmission oil level low	Refill
Gears worn, scuffed, or broken	Replace
Countershaft sprung or badly worn	Replace
Excessive end play of countershaft	Adjust or replace worn parts
TRANSMISSION NOISY WHILE IN GEAR	
Transmission not aligned with engine	Align
Bearings dry, badly worn, or broken	Lubricate or replace
Transmission oil level low	Refill
Gears worn, scuffed, or broken	Replace
Countershaft sprung or badly worn	Replace
Excessive end play of countershaft	Adjust or replace worn parts
Main shaft rear bearing worn or broken	Replace
Gear teeth worn	Replace gear
Engine vibration dampener defective	Replace or adjust
Speedometer drive gears worn	Replace
Clutch friction disk defective	Replace
Gears loose on main shaft	Replace worn parts
Transmission oil level low	Refill
TRANSMISSION HARD TO SHIFT	
Engine clutch not releasing	Adjust
Sliding gear tight on shaft splines	Clean splines or replace shaft or gear
Shift linkage out of adjustment	Adjust
Main shaft splines distorted	Replace or clean splines
Synchronizing unit damaged	Replace defective parts
Sliding gear teeth damaged	Replace
GEARS CLASH WHEN SHIFTING	
Clutch not releasing	Adjust
Synchronizer unit defective	Replace defective parts
Gears sticky on main shaft	Free up gears. Replace defective gears
TRANSMISSION STICKS IN GEAR	
Clutch not releasing	Adjust
Detent balls stuck	Free
Shift linkage out of adjustment or needs lubricating	Adjust or lubricate
Sliding gears tight on shaft splines	Clean splines or replace shaft or gears
TRANSMISSION SLIPS OUT OF GEAR	
Shift linkage out of adjustment	Adjust
Gear loose on shaft	Replace shaft or gear
Gear teeth worn	Replace gear
Excessive end play in gears	Replace worn or loose parts
Lack of spring tension on shift lever detent	Install new spring
Badly worn transmission bearings	Replace

Continued on next page

MANUAL TRANSMISSIONS

MANUAL TRANSMISSION TROUBLESHOOTING	
POSSIBLE CAUSE	REMEDY
TRANSMISSION LEAKS OIL	
Oil level too high	Drain to proper level
Gaskets damaged or missing	Install new gaskets
Oil seals damaged or improperly installed	Install new oil seals
Oil throw rings damaged, improperly installed, or missing	Install oil throw rings properly or replace
Drain plug loose	Tighten
Transmission case bolts loose, missing, or threads stripped	Tighten or replace
Transmission case cracked	Replace
Lubricant foaming excessively	Use recommended lubricant

TEST YOURSELF

QUESTIONS

1. What are the three types of mechanisms used to engage gears?

2. (True or False?) In some sliding gear transmissions, the input and output shafts are locked together in high gear.

3. What is one advantage of the collar shift transmission over the sliding gear?

4. Which type of transmission allows on-the-go shifting while the machine's wheels are still rolling?

5. What is the purpose of the countershaft?

6. What are the three functions of a synchronizer?

POWER SHIFT TRANSMISSIONS

INTRODUCTION

A power shift transmission performs exactly the same functions as a manual transmission, which was discussed in Chapter 3. At a glance, you would hardly notice the difference between them.

Indeed, the distinction is somewhat of a gray area. Definitions vary. Perhaps the best way to differentiate them is this:

- In a manual transmission, the operator uses mechanical linkage to engage the gears by manual force.
- In a power shift transmission, the machine's power is used to engage the gears, usually through multiple hydraulically engaged clutches.

Most power shift transmissions can be shifted on-the-go without using the clutch pedal and without any interruption of power. Most, but not all, can be started and stopped without using the clutch pedal. Transmission controls are usually convenient to reach and easy to move. Although there are other benefits, the hallmark of power shift is convenience.

Many transmissions are hybrids — partly manual and partly power shift. Perhaps the most common examples use manual shift for ranges and power shift for multiple speeds within each range. These are sometimes billed as power shift transmissions, but a more accurate term would be partial power shift.

Continued on next page

POWER SHIFT TRANSMISSIONS

More than 50 years ago, certain manufacturers began offering an overdrive, which could be shifted on-the-go without clutching. This doubled the number of speeds and provided a handy way to downshift for headlands or tough spots. Was this a power shift? Technically no, if the operator was engaging a clutch through mechanical linkage, but it was the forerunner of many designs to come.

Before long, true partial power shift transmissions were available in the form of a hydraulically operated hi-lo (Fig. 1). All the operator did was flip a lever to shuttle a valve. Hydraulic pistons did the shifting.

Fig. 1 — Hydraulic Hi-Lo, an Early Partial Power Shift Transmission

For loader work, a hydraulic reverser (Fig. 2) is about the handiest feature ever invented. A fingertip lever is all you need to shuttle between forward and reverse. It has automatic modulation for smooth engagement in both directions. After decades on the market, hydraulic reversers are still extremely popular today. The rest of the transmission is usually manual.

A—Forward   B—Reverse

Fig. 2 — Hydraulic Reverser

ADDITIONAL BENEFITS

Power shift controls offer more than just convenience. Operation is simpler. An inexperienced operator can learn more quickly and make fewer mistakes. A power shift transmission is more nearly foolproof and abuse-proof. Any operator can be more productive, adjusting speed to match changing conditions.

Power shift controls are far more adaptable for adding features:

- Safety features to protect the operator.
- Default features to protect the machine.
- Automatic modulation for smooth starts.
- Automatic modulation for smooth shifts.
- Self-calibration.
- Self-diagnosis.
- Automatic shifting.

Keep in mind that a power shift is not infinitely variable like a hydrostatic or infinitely variable transmission (IVT). It has a set number of speeds, just like a manual transmission, but the power shift is much more flexible for selecting the one you want.

POWER SHIFT TRANSMISSIONS

HYDRAULIC CLUTCHES

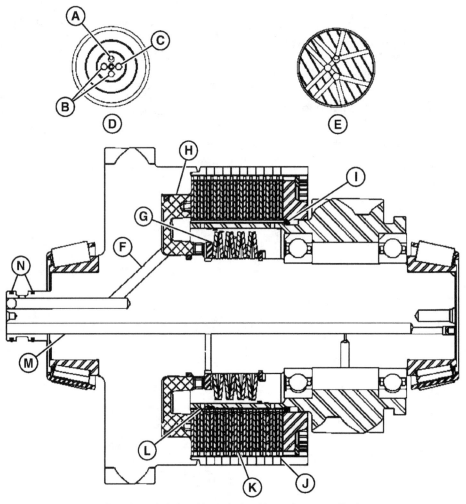

Fig. 3 — Hydraulic Clutch for Locking a Gear to a Shaft

A—Lube Passage
B—Lube Passage
C—System Pressure Passage
D—Seal Ring End of Shaft
E—Cross-Section of Shaft Assembly
F—System Pressure Passage
G—Shut-Off Washer
H—Clutch Piston
I—Sealing Ring
J—Oil Return to Sump
K—Clutch and Separator Plates
L—Passage to Lubricate Clutch and Separator Plates
M—Lube Passage
N—Seal Rings

Remember from chapter 3 that the only difference between collar shift and synchronizer is the device that locks the gear to the shaft? Power shift provides yet another way to do the same thing. Although it hasn't been done, it would be possible to design a single transmission with three options for gear engagement — shift collars, synchronizers, or hydraulic clutches.

Fig. 3 shows one shaft from a power shift transmission. It has a gear at each end. The large one on the left is part of the shaft. The small one on the right floats on ball bearings. Engaging a hydraulic clutch locks the two together. The clutch hub is attached to the floating gear and splined to 14 clutch disks. The clutch drum is attached to the shaft and splined to 14 separator plates.

To engage the clutch, we squeeze disks and plates together by applying oil pressure behind the piston. Oil is routed through passages drilled in the shaft. Sealing rings prevent leakage between shaft and housing.

To release the clutch, we shut off the pressure and dump it to sump. The piston is pushed back by a stack of spring washers.

Notice the additional oil passages drilled in the shaft. They supply oil to lubricate and cool the clutch and bearings. Lube pressure is much lower than control pressure.

Continued on next page

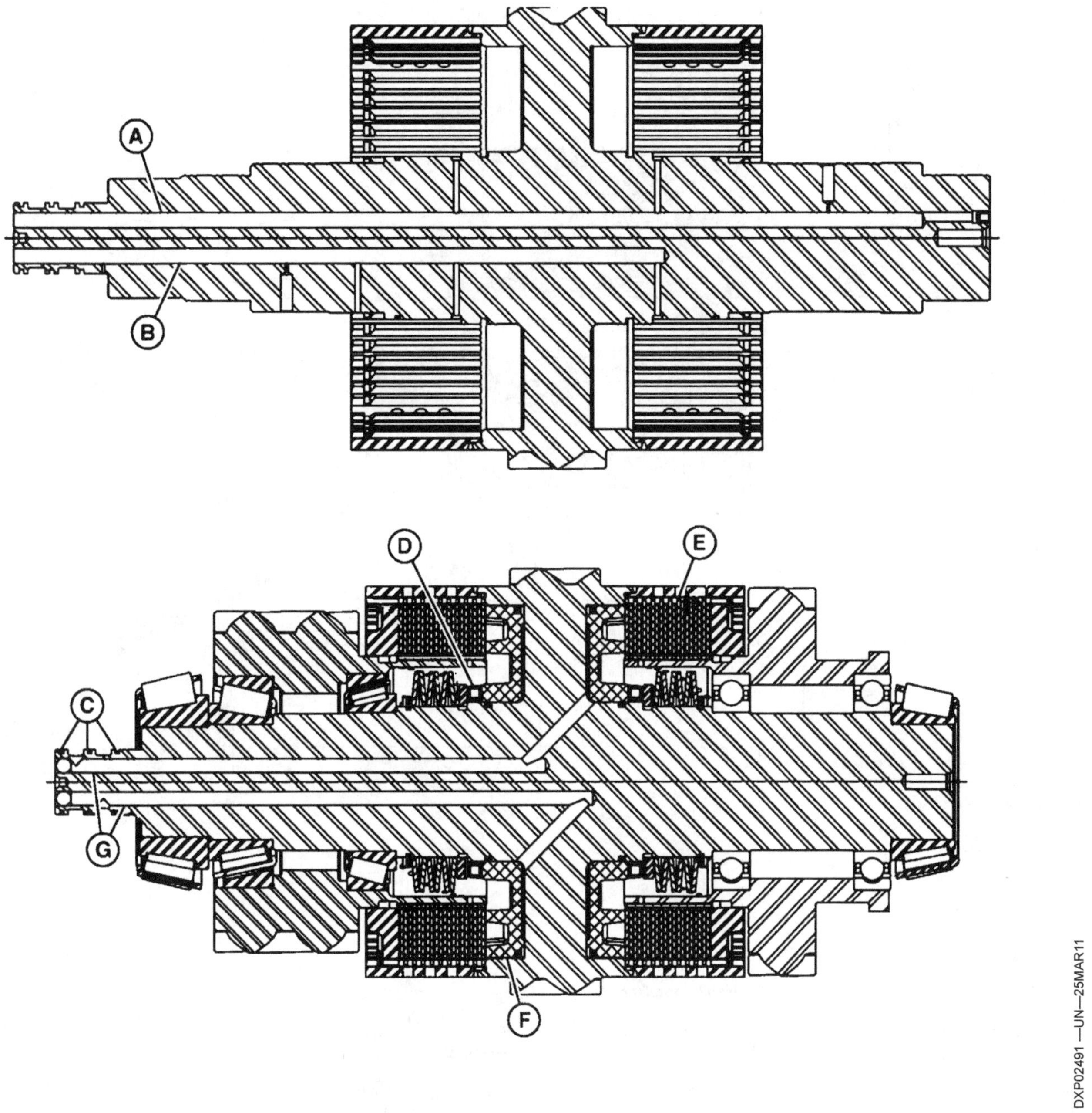

POWER SHIFT TRANSMISSIONS

Fig. 4 — Shaft with Two Floating Gears and Two Hydraulic Clutches

A—Lube Passage
B—Lube Passage
C—Seal Rings
D—Lube Shut-Off Washer
E—Clutch and Separator Plates
F—Piston
G—System Pressure Passage

Fig. 4 shows another shaft from the same transmission. This one has two floating gears, two hydraulic clutches, and two control oil pressure passages. A third sealing ring isolates each oil passage from the other. The clutches are engaged and released the same way as in the previous example. This shaft apparently has less torque, as these clutches use only 10 disks and plates each.

POWER SHIFT TRANSMISSIONS

HYDRAULIC REVERSER

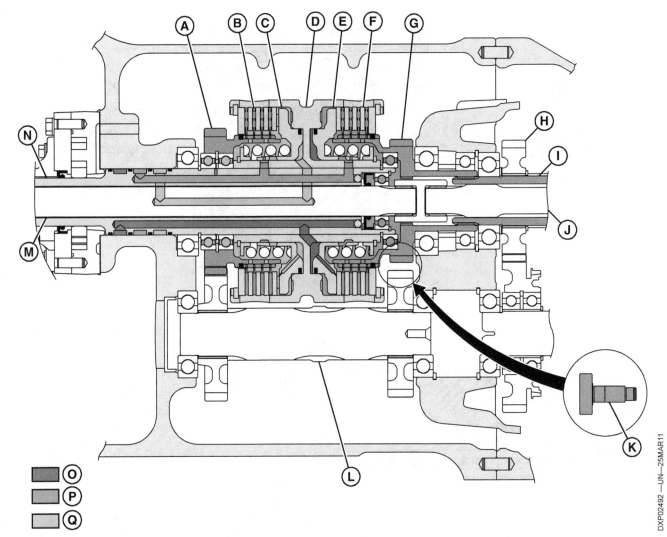

Fig. 5 — Hydraulic Reverser in Forward

A—Reverse Drive Gear
B—Reverse Plates and Disks
C—Reverse Piston
D—Clutch Drum
E—Forward Piston
F—Forward Plates and Disks
G—Forward Drive Gear
H—Transmission Second-Speed Drive Gear
I—Transmission Top Shaft
J—PTO Shaft
K—Reverse Idler Gear
L—Countershaft Speed Sensor Teeth
M—PTO Drive Shaft
N—Traction Drive Shaft
O—High Pressure Oil
P—Lubrication Oil
Q—Return Oil

Fig. 5 shows the front section of a partial power shift transmission. Out of sight to the right is a 12-speed manual transmission. What we see here is a power shift reverser on the input shaft.

The top shaft assembly is similar to what we saw in Fig. 4, except for one detail. Although the forward drive gear is floating on the first shaft, it is splined to the transmission top shaft, which is the output from the reverser and input to the main transmission.

In this illustration, the forward clutch is engaged. The clutch drum is splined to the traction drive shaft, so now the forward drive gear and transmission top shaft are also tied to it. The whole assembly rotates as one piece.

Continued on next page

POWER SHIFT TRANSMISSIONS

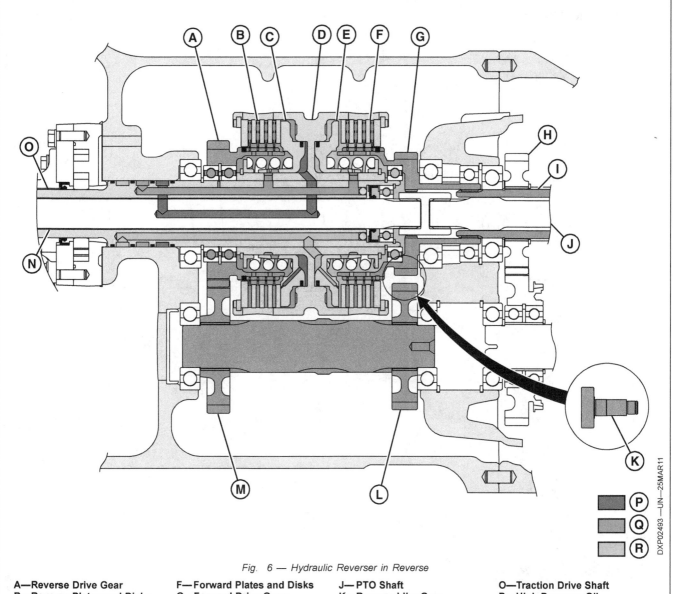

Fig. 6 — Hydraulic Reverser in Reverse

A—Reverse Drive Gear
B—Reverse Plates and Disks
C—Reverse Piston
D—Clutch Drum
E—Forward Piston
F—Forward Plates and Disks
G—Forward Drive Gear
H—Transmission Second-Speed Drive Gear
I—Transmission Top Shaft
J—PTO Shaft
K—Reverse Idler Gear
L—Forward Driven Gear
M—Reverse Driven Gear
N—PTO Drive Shaft
O—Traction Drive Shaft
P—High Pressure Oil
Q—Lubrication Oil
R—Return Oil

Now let's flip the control lever to reverse (Fig. 6). We release the forward clutch and engage the reverse clutch. This locks the reverse drive gear onto the traction drive shaft. The new power flow is down to the countershaft through the front gears. The rear gear on the countershaft drives a reverse idler, which turns the forward drive gear in the opposite direction.

One clutch turns the transmission top shaft forward; the other clutch turns it backward.

POWER SHIFT TRANSMISSIONS

HYDRAULIC HI-LO

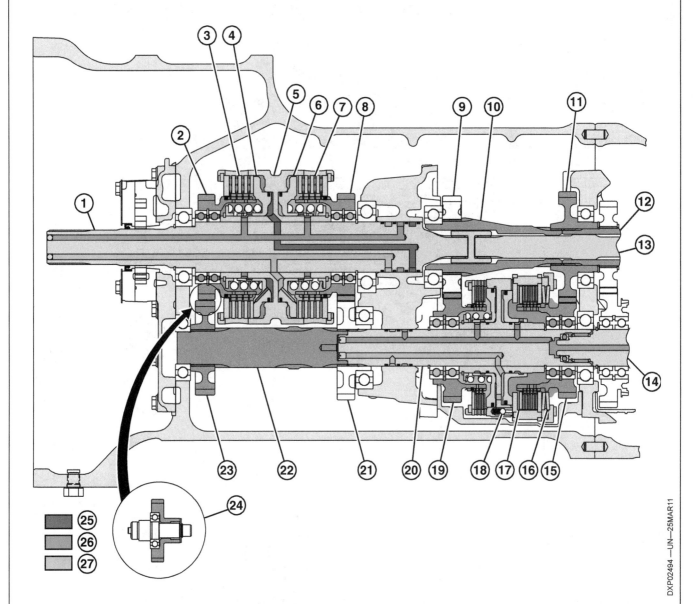

Fig. 7 — Hydraulic Hi-Lo in Low

1— Traction Drive Shaft
2— Reverse Drive Gear
3— Reverse Plates and Disks
4— Reverse Piston
5— Clutch Drum
6— Forward Piston
7— Forward Plates and Disks
8— Forward Drive Gear
9— High-Speed Driven Gear
10— Connect Shaft
11— Low-Speed Driven Gear
12— Transmission Top Shaft
13— PTO Shaft
14— Transmission Bottom Shaft
15— Low-Speed Drive Gear
16— Low-Side Engagement Springs
17— Clutch Plate
18— Bleeder
19— High-Speed Drive Gear
20— Hi-Lo Clutch Shaft
21— Forward Driven Gear
22— Countershaft
23— Reverse Driven Gear
24— Reverse Idler Gear
25— High Pressure Oil
26— Pressurized Lubrication Oil
27— Low Pressure/Return Oil

This versatile partial power shift transmission can add yet another option. Fig. 7 shows it equipped with both a hydraulic reverser and a hydraulic hi-lo.

Unfortunately for the sake of comparison, the reverser configuration is different on machines that also have hi-lo.

The reverse idler gear is moved to the front, and it is the bottom shaft that delivers power to the next stage.

For now, let's concentrate on the hi-lo, which is the right-hand portion of the diagram.

Continued on next page

POWER SHIFT TRANSMISSIONS

The hi-lo clutch shaft delivers power to the hi-lo clutch assembly. The high- and low-speed drive gears float on the shaft. Each of these gears has a clutch to lock it to the shaft.

But look closely at the low-speed clutch. This one is different from any of the others shown so far. It has spring engagement and pressure release. Two low-side engagement springs press the disks and plates against the clutch plate. These are very strong spring washers. With no hydraulic pressure applied, the low-speed clutch is engaged.

Continued on next page

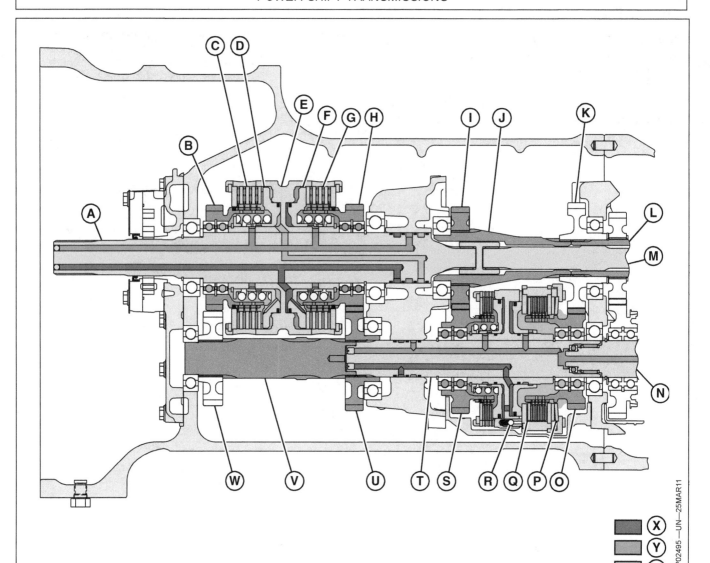

Fig. 8 — Hydraulic Hi-Lo in High

A—Traction Drive Shaft
B—Reverse Drive Gear
C—Reverse Plates and Disks
D—Reverse Piston
E—Clutch Drum
F—Forward Piston
G—Forward Plates and Disks
H—Forward Drive Gear
I— High-Speed Driven Gear
J—Connect Shaft
K—Low-Speed Driven Gear
L—Transmission Top Shaft
M—PTO Shaft
N—Transmission Bottom Shaft
O—Low-Speed Drive Gear
P—Low-Side Engagement Springs
Q—Clutch Plate
R—Bleeder
S—High-Speed Drive Gear
T—Hi-Lo Clutch Shaft
U—Forward Driven Gear
V—Countershaft
W—Reverse Driven Gear
X—High Pressure Oil
Y—Pressurized Lubrication Oil
Z—Low Pressure/Return Oil

When pressure oil is directed to the piston, it releases the clutch. This is difficult to see in the diagram. The clutch plate doesn't move. Instead, pins at the outside push a plate at the rear, which compresses the low-side engagement springs to release the disks and plates.

Notice the oil passage between the two clutches. One valve sends oil to both clutch pistons. This releases the low clutch and engages the high clutch.

Why? Because a spring-engaged clutch will not slip if control pressure is low. If system pressure drops low enough to risk slipping, the control valve automatically shifts to low and prevents damage. Several transmission designs incorporate this feature.

POWER SHIFT TRANSMISSIONS

PLANETARY PARTIAL POWER SHIFT

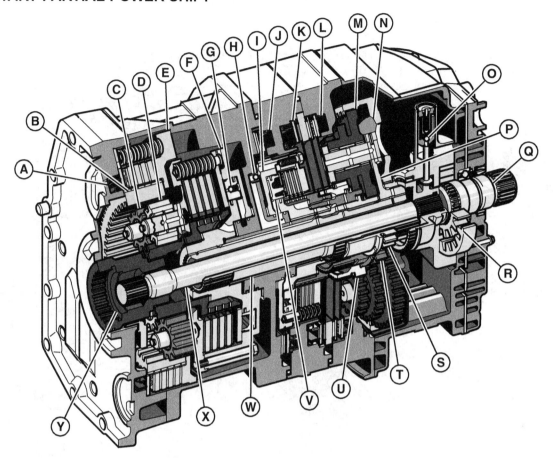

Fig. 9 — Partial Power Shift Transmission with Planetary Gears

A—Reverse Brake Piston
B—Ring Gear
C—Outer Planetary Gear
D—Inner Planetary Gear
E—Pressure Plate with Disk Pack, Reverse Brake
F—Forward Clutch with Planetary Carrier, Reverse Brake
G—Piston with Overspeed Relief Valve, Forward Clutch
H—Suction Valve
I—C4 Clutch Drum
J—3rd Gear Piston
K—2nd Gear Piston
L—1st Gear Piston
M—Planet Pinion Gear
N—Ring Gear
O—Pneumatic Pump
P—Planet Pinion Carrier
Q—Transmission Input Shaft
R—Transmission Oil Pump
S—3rd Speed Sun Gear
T—2nd Speed Sun Gear
U—1st Speed Sun Gear
V—Piston with Anti-Cavitation Check Valve C4
W—Drive Shaft
X—PTO Driveline
Y—Transmission Output Shaft

Here's the power shift portion of a more sophisticated transmission (Fig. 9). It is attached to the front of a synchronized range box with four, five, or six ranges. This module contains a reverser and four-speed power shift assembly. In combination with the range box, it provides 16, 20, or 24 speeds in both directions. (Control linkage and common sense may prevent engaging the fastest reverse speeds, but the transmission is capable of producing them.)

This example uses planetary gears. There are no countershafts until you get to the range box. Planetaries are compact, efficient, and versatile, but the power flows are harder to trace.

The transmission input shaft (Q) is on the right-hand side. It turns the ring gear (N) for a compound planetary that produces four speeds. The planet pinions (M) are one-piece triple gears. There are three sun gears (S, T, and U). The planet pinion carrier (P) is the output. It is connected to a drive shaft (W) that carries power to the reverser.

The reverser contains a forward clutch (F), reverse brake (E), and planetary gear set (B, C, and D). The transmission output shaft (Y) goes into the range box.

Continued on next page

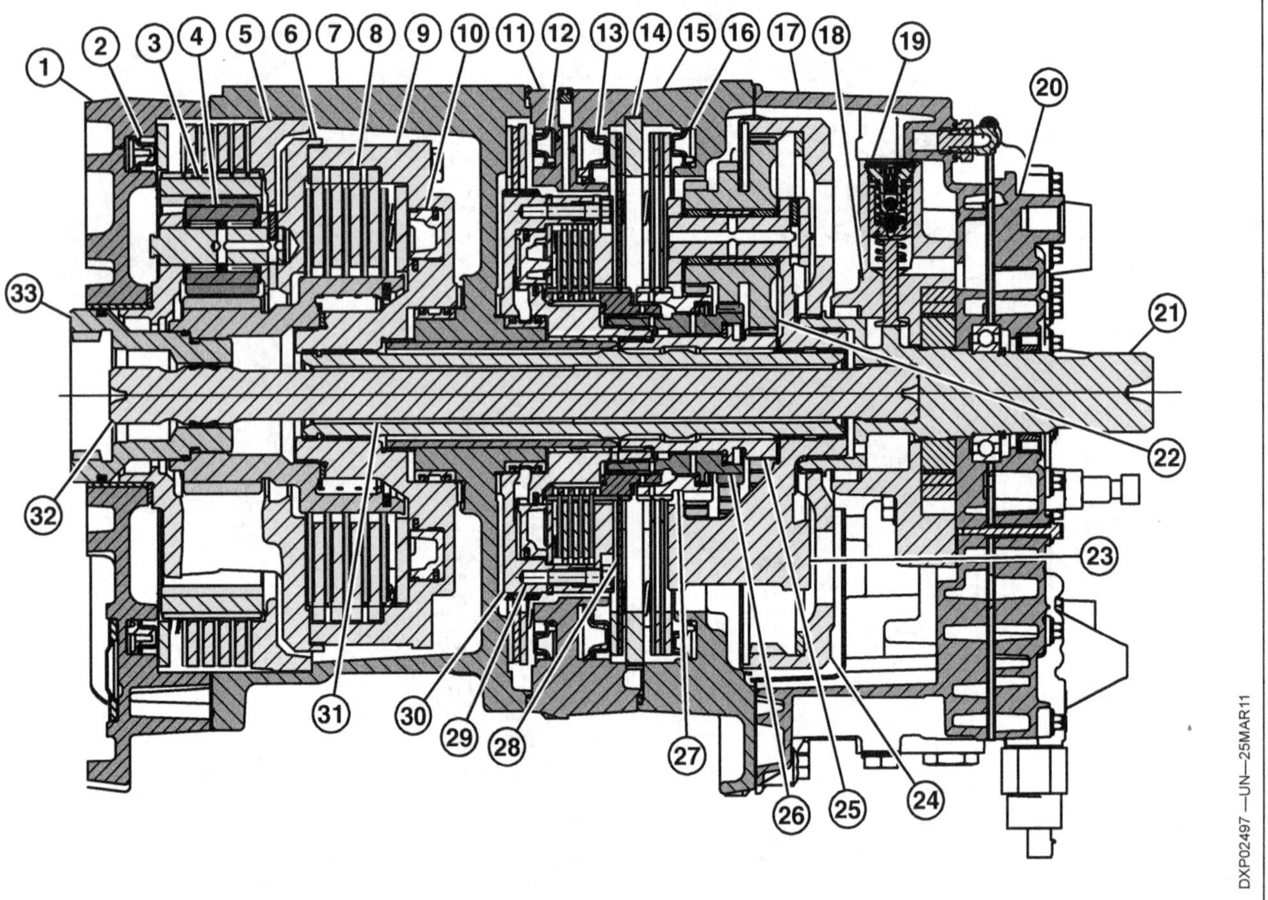

POWER SHIFT TRANSMISSIONS

Fig. 10 — Cross-Section of Partial Power Shift Transmission

1— Reverse Brake Housing
2— Reverse Brake Piston
3— Planetary Ring Gear
4— Planetary Pinion Gear
5— Pressure Plate with Disk Pack, Reverse Brake
6— Forward Clutch with Planetary Carrier, Reverse Brake
7— Planetary Gear Housing and C4 Clutch
8— Clutch Hub
9— Forward Clutch
10— Forward Clutch Piston
11— B2-B3 Brake Housing
12— 3rd Gear Piston Brake
13— 2nd Gear Piston Brake
14— Separator Plate
15— B1 Brake Housing
16— 1st Gear Piston Brake
17— Front Transmission Cover
18— Transmission Oil Pump
19— Pneumatic Pump
20— Front Valve Housing
21— Transmission Input Shaft
22— Planetary Pinion Gear
23— Planetary Carrier
24— Planetary Ring Gear
25— 3rd Gear Sun Gear
26— 2nd Gear Sun Gear
27— 1st Gear Sun Gear
28— Clutch Hub
29— C4 Clutch Piston
30— C4 Clutch Drum
31— Drive Shaft
32— PTO Driveline
33— Transmission Output Shaft

To see how it operates, let's use this straight cross-section diagram (Fig. 10). Sometimes it helps to refer to both views.

The transmission input shaft (21) turns engine speed at all times. It operates the transmission oil pump (18) and pneumatic pump (19). It is splined to the PTO driveline (32), which has nothing to do with transmission operation. What is important is that the input shaft is also splined to the planetary ring gear (24). So the ring gear turns engine speed at all times. The planetary pinion gears (22) are triple gears. There are three sun gears (25, 26 and 27). Each sun gear has its own piston brake (12, 13 and 16).

By engaging one brake and thus locking one sun gear stationary, we force the planet pinions to walk around the sun and turn the planet pinion carrier (23). Each sun gear produces a different reduction ratio.

The fourth speed is produced by locking two sun gears together. The planet pinions can not rotate, because they are splined to two gears with different ratios that are locked together. This forces the entire assembly—ring gear, planet pinion carrier, and all three sun gears—to revolve as one piece. This is direct drive. To get it we engage C4 clutch (29), which locks 2nd speed sun gear (26) and 3rd speed sun gear (25) together.

Continued on next page

POWER SHIFT TRANSMISSIONS

The planet pinion carrier (23) is the output. It is splined to the drive shaft (31). So is the forward clutch drum (9). If the forward clutch is engaged, the clutch hub (8) turns the output shaft (33). The forward clutch drum is bolted to the planet pinion carrier (6) for the reverse planetary.

The forward clutch hub is also the sun gear for the reverse planetary. If the forward clutch is released and the reverse brake (2) is engaged, the planetary ring gear (3) is held stationary. The planetary pinion gears (4) walk around the ring gear and turn the sun gear.

This is a reversing planetary (Fig. 11) because the planet pinions are in pairs, with one contacting the ring gear and the other contacting the sun gear.

The reverser also functions as the traction clutch. When you push the clutch pedal, it releases the forward clutch or reverse brake, whichever was engaged. The same was true in our previous example, which used a countershaft instead of planetary gears.

A—Planet Pinions
B—Sun Gear
C—Ring Gear

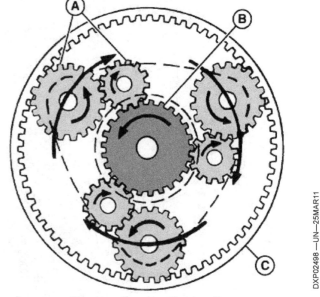

Fig. 11 — Reversing Planetary Gears

POWER SHIFT TRANSMISSIONS

FULL POWER SHIFT

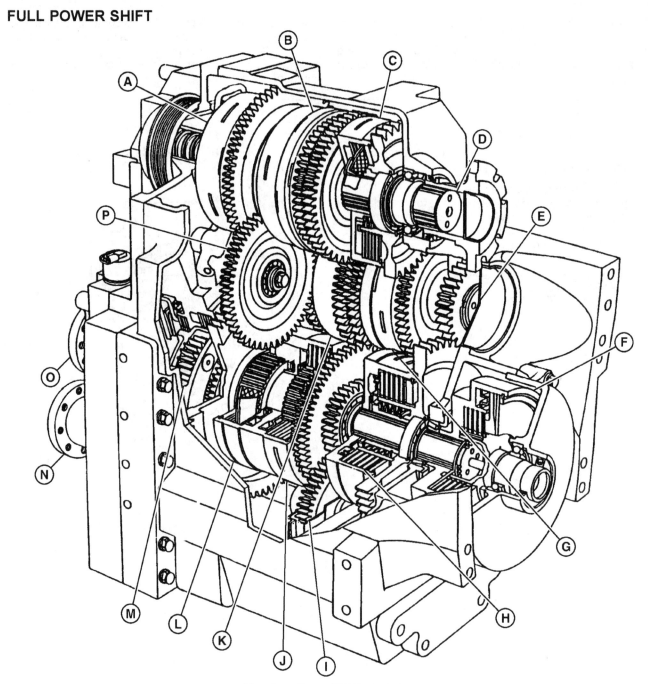

Fig. 12 — Power Shift Transmission

A—C4 (Clutch 4)
B—C1 (Clutch 1)
C—C3 (Clutch 3)
D—Input Shaft from Engine
E—Countershaft
F—MFWD Clutch
G—C2 (Clutch 2)
H—BC (B Clutch)
I—Auxiliary Drive Gear
J—DC (D Clutch)
K—CR (Reverse Clutch)
L—CC (C Clutch)
M—AB (A Brake)
N—PTO Drive Shaft
O—Output Shaft
P—Reverse Idle Gear

The previous examples were all partial power shift. Many transmissions are full power shift, with no manual components. Fig. 12 shows a power shift transmission for large tractors. It has 16 forward speeds and 4 reverse. It has three shafts plus a reverse idler. Notice the large vertical "drop" between the input shaft and the output shaft. There is also a PTO drive shaft, which isn't involved in transmission operation.

Continued on next page

POWER SHIFT TRANSMISSIONS

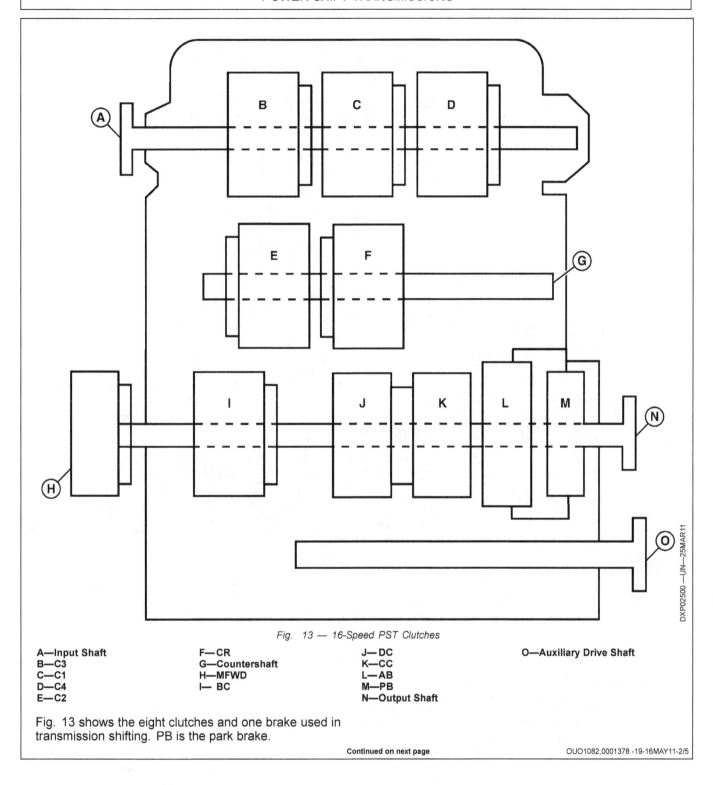

Fig. 13 — 16-Speed PST Clutches

A—Input Shaft
B—C3
C—C1
D—C4
E—C2
F—CR
G—Countershaft
H—MFWD
I—BC
J—DC
K—CC
L—AB
M—PB
N—Output Shaft
O—Auxiliary Drive Shaft

Fig. 13 shows the eight clutches and one brake used in transmission shifting. PB is the park brake.

Continued on next page

POWER SHIFT TRANSMISSIONS

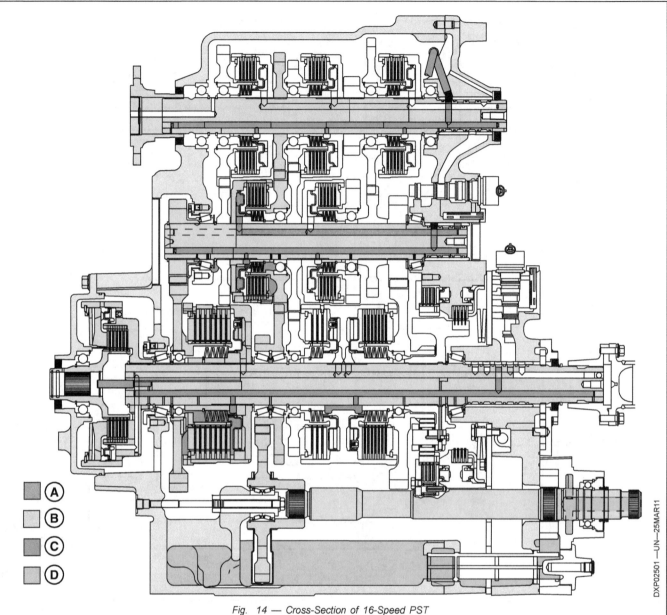

Fig. 14 — Cross-Section of 16-Speed PST

A—System 1 Oil
B—System 2 Oil
C—Lube Oil
D—Sump Oil

Here is a cross-section of that transmission. It contains only one planetary gear set, for A Range on the bottom shaft. All the others use floating gears and hydraulic clutches. It is shown in 7th speed, which engages C2 and BC clutches. Given an element engagement chart, you could easily trace the power flow in any speed. Each speed uses one of the speed clutches (C1, C2, C3, C4, or CR) and one of the ranges (AB, BC, CC, or DC).

The park brake is spring engaged. Strong spring washers squeeze the disks and plates to hold the A Range planet pinion carrier, which prevents the output shaft from turning. The machine cannot move until pressure oil is directed to the park brake release piston.

The MFWD clutch is used only on machines with mechanical front wheel drive. It is also spring engaged.

The PTO drive shaft turns all the time. Another clutch at the rear of the machine controls engagement of PTO implements.

Continued on next page

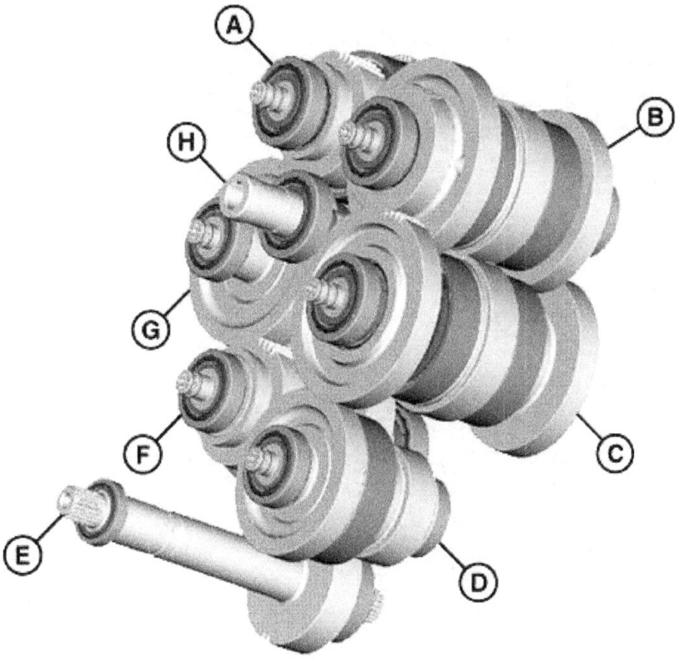

POWER SHIFT TRANSMISSIONS

Fig. 15 — 18-Speed PST for Largest Four-Wheel-Drive Machines

A—Second Stage Shaft Assembly (A, 2)
B—Third Stage Shaft Assembly (B, 1)
C—Fourth Stage Shaft Assembly (C, R)
D—Fifth Stage Shaft Assembly (L)
E—Output Shaft Assembly
F—Sixth Stage Shaft Assembly (H)
G—Seventh Stage Shaft Assembly (M)
H—First Stage (Input) Shaft Assembly

Fig. 15 shows an even larger power shift transmission. It provides 18 speeds forward and 6 reverse. It has eight shafts and nine clutches. The clutches are classified as:

- Three direction clutches (1, 2, R)
- Three speed clutches (A, B, C)
- Three range clutches (L, M, H)

Every transmission speed uses one clutch of each type. The following element engagement chart shows which ones, plus the overall gear ratios.

Continued on next page

18-Speed PST Gear Table		
Gear	Clutches Engaged	Gear Ratio
1F	1AL	4.648
2F	2AL	3.776
3F	1AM	3.410
4F	1BL	3.054
5F	2AM	2.771
6F	2BL	2.481
7F	1BM	2.241
8F	1CL	2.007
9F	2BM	1.821
10F	2CL	1.631
11F	1CM	1.473
12F	1AH	1.333
13F	2CM	1.197
14F	2AH	1.083
15F	1BH	0.876
16F	2BH	0.712
17F	1CH	0.576
18F	2CH	0.468
N	—	—
1R	RAL	4.648
2R	RAM	3.410
3R	RBL	3.054
4R	RBM	2.241
5R	RCL	2.007
6R	RCM	1.473

PLANETARY POWER SHIFT

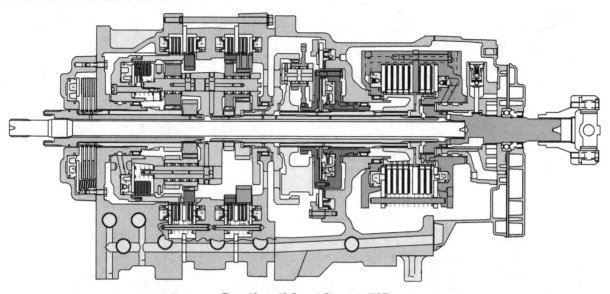

Fig. 16 — 19-Speed Planetary PST

Finally, here is a 19-speed power shift transmission using all planetary gears. It has four clutches and six brakes. It has one compound planetary after another. Rather than locking one element of a planetary gear set, it often drives two inputs at different speeds. It is, of course, possible to trace power flow in any speed, but very time-consuming and confusing.

This basic design began as an 8-speed. It was expanded to 15 speeds and eventually to 19. Altogether, it was in production about 40 years.

More combinations are available than the 19 we use. Some are so nearly the same speed, there is no point in using both. This is similar to the "13-speed" transmission so popular in heavy trucks, which actually would have 20 speeds if we used all of them.

OPERATOR CONTROLS

Transmission controls vary widely. No two manufacturers offer quite the same options, and no two options use quite the same controls. Here are a few examples.

Remember the four-speed planetary in Figs. 9 and 10? Fig. 17 shows the standard control lever. It has four positions in a straight line. It can be shifted on-the-go without using the clutch pedal. The lever beside it is the range shifter. Ranges are synchronized for shifting on-the-go, but you must use the clutch as in a car or truck.

A—Standard Control Lever

Fig. 17 — Partial Power Shift Controls, Basic

Continued on next page

That transmission also included a reverser. Fig. 18 shows the reverser control lever. It has three positions — forward, neutral, and reverse. You can shuttle between forward and reverse without using the clutch pedal. The reverser has automatic modulation for smooth engagement.

You can also use the reverser control lever to start and stop the machine. The lever must be in neutral when starting the engine.

A—Reverse Drive Lever in Forward Position
B—Reverse Drive Lever in Neutral Position
C—Reverse Drive Lever in Reverse Position
D—Reverser Control

Fig. 18 — Reverser Control Lever

Want more sophistication? Fig. 19 shows the same partial power shift with all the options. It has six ranges instead of four, and speeds within a range are shifted electrohydraulically. You just push a button to shift up or down.

This version also has "speed matching." Whenever you engage the clutch, it automatically selects the speed that most nearly matches the speed you're moving.

Suppose the machine is parked. You start the engine, put it in C Range, and flip the reverser control lever to forward. It will automatically start in C-1. Now you push the upshift button to shift into C-2, C-3, C-4. Now you push the clutch pedal and shift to D Range. It will automatically select D-2, because that's closest to the speed you're moving.

Or suppose you're driving in D-4 and slowing down for a corner. You could push the downshift button to shift to D-3, D-2, D-1. Or you could just push the clutch pedal and coast down to the slower speed. When you re-engage the clutch, it will be in D-1.

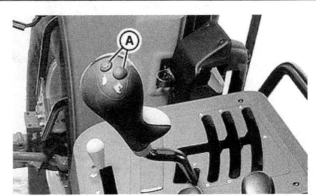

Fig. 19 — Partial Power Shift Controls, Optional

A—Range Shift Buttons

Continued on next page

POWER SHIFT TRANSMISSIONS

Fig. 20 shows a common type of power shift control lever. You can engage any speed at any time. You can shift on-the-go without using the clutch pedal. You can start and stop or shuttle between forward and reverse without using the clutch pedal. This lever controls the valves (through linkage or electrohydraulics) that engage the transmission elements.

Fig. 20 — Power Shift Transmission Control

For fingertip convenience, Fig. 21 shows the power shift control is built into the armrest on the seat. It has slots for forward, reverse, and park. Within the forward and reverse slots, you can shift up or down by bumping the lever. Again, you can start, stop, shift, or shuttle without using the clutch pedal. It has automatic modulation and speed matching.

This transmission is computer controlled. It senses travel speed and engine load to adjust modulation. It allows self-calibration and self-diagnosis. It can automatically downshift if the engine is overloaded. It is programmable within limits. For instance, you can tell it which gear to engage when the lever goes into the forward or reverse slot. (There's no need to start in 1st each time if the job doesn't require it.)

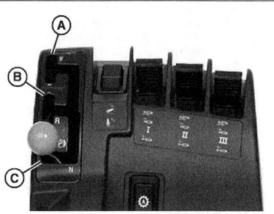

Fig. 21 — Armrest Control

A—Front Slot (Forward)
B—Center Slot (Reverse)
C—Rear Slot (Park)

COMPUTER CONTROLS

Integrated logic has revolutionized power shift transmission controls (Fig. 22). A transmission today may use more computer technology than the satellites that carried men to the moon. Far more. The applications are limitless:

- Engagement override to prevent movement if the engine is jump-started in gear.
- Self-calibration to maintain optimum shift quality.
- Load and speed sensing to automatically adjust shift characteristics to conditions.
- Speed matching.
- Default modes. If one transmission element is defective, the controller can default to another speed to prevent damage. It can default to neutral if operation would risk machine damage or personal injury.
- "Come-home" mode to allow the machine to be driven slowly in one gear when normal operation isn't possible. This enables you to load the machine onto a truck or drive it to the shop for repairs.
- An unprecedented array of convenience features.

Automatic shifting is increasingly popular. This isn't equivalent to an automatic transmission in a car, but it

Fig. 22 — Power Shift Transmission Controller

enables the machine to downshift and upshift as load changes.

Computers also enable the operator to program the machine to perform multiple functions at the touch of a button. Fig. 23 shows a popular version that can include transmission, hitch, remote hydraulic, PTO, and MFWD functions. The operator selects whichever ones are useful.

For example, it could be programmed for headland turns at the ends of a farm field. When the tractor reaches the end of the field, the operator can push one button to downshift twice, raise the hitch, extend a remote hydraulic cylinder, shut off PTO, and disengage MFWD. After completing the turn, the operator can push a second button to return everything to the previous settings.

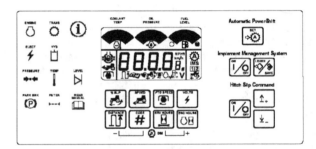

Fig. 23 — Programmable Multi-Function Control

POWER SHIFT TRANSMISSIONS

CONTROL VALVES

Fig. 24 identifies 18 control valves on a power shift transmission housing, plus an assortment of pressure sensors and test ports. Most of the valves control hydraulic clutches for element engagement. The rest are system valves for pressure regulation, lube relief, cooler relief, etc.

A control valve must supply pressure oil to its clutch as needed. It must shut off incoming oil and dump pressure from the element as needed. Engagement oil is routed through drilled passages in manifolds and shafts.

A few valves are directly controlled by the operator moving a lever. Others are pilot operated, meaning a small volume of oil from another valve moves the flow control valve. Most transmission control valves are now operated electronically, often by a computer.

Many valves are ON-OFF, open or closed, with no in-between. Many have a detent for each position, with snap action to prevent ever stopping between positions.

1— Input Shaft
2— CR Clutch Solenoid Valve
3— C1 Test Port
4— C1 Clutch Solenoid Valve
5— C2 Clutch Solenoid Valve
6— C3 Clutch Solenoid Valve
7— C4 Clutch Solenoid Valve
8— C4 Test Port
9— C3 Test Port
10— Countershaft—Behind Manifold
11— CR Test Port
12— Park Brake Release
13— Park Brake Release
14— Park Brake Pressure Sensor (Test Port)
15— A-Brake Solenoid Valve
16— C2 Test Port
17— B-Clutch Solenoid Valve
18— C-Clutch Solenoid Valve
19— D-Clutch Solenoid Valve
20— MFWD Test Port
21— Lube Cutoff Valve
22— D-Clutch Test Port
23— Vent
24— C-Clutch Test Port
25— B-Clutch Test Port
26— Clutch Pressure Sensor
27— Brake Valve Sump
28— A-Brake Test Port
29— Park Brake Test Port (Pressure Port)
30— Park Brake Release Pump
31— Clutch Disengage Switch
32— Ground Speed Park Control Valve
33— GDP Inlet Passage Plug
34— GDP 5-Bar Relief Valve
35— Clutch Outer Arm
36— Ground Driven Pump
37— Output Shaft
38— Cooler Relief Valve
39— Lube Relief Valve
40— Transmission Lube
41— Shift Accumulator (Internal)
42— Scavenge Pump
43— Transmission Inlet
44— Steering Load Sense
45— Emergency Steering Valve
46— Steering Pilot Oil
47— GDP 60-Bar Relief Valve
48— Auxiliary Output Shaft
49— Sump Screen
50— Pressure Regulating Valve
51— Oil Cooler

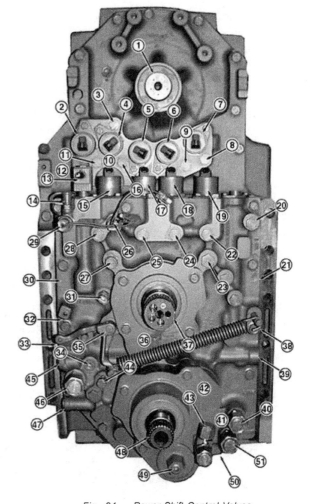

Fig. 24 — Power Shift Control Valves

Continued on next page

POWER SHIFT TRANSMISSIONS

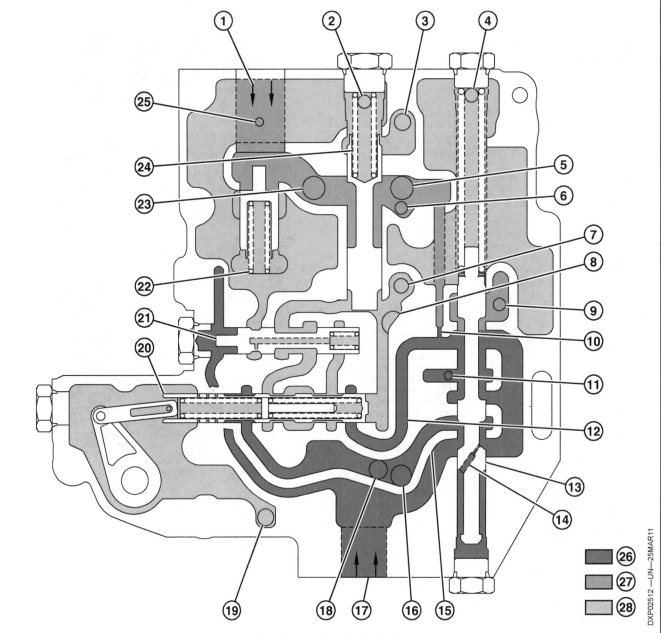

Fig. 25 — Spring-Type Modulating Valve

1— From Oil Cooler
2— Sump Passage
3— Passage to Clutch Lube
4— Sump Passage
5— Lube Pressure DR
6— Clutch Pressure DR
7— Passage to Traction Clutch
8— Clutch Pressure DR
9— Port to Oil Cooler
10— Orifice to Priority Lube
11— System Pressure Sender
12— To Clutch Valve
13— Pressure Regulating Valve
14— Orifice
15— Engagement Pressure Passage
16— Passage to Two Speed
17— From Transmission Filter and Pump
18— Port to PTO Clutch and Differential Lock
19— Clutch Valve Sump
20— Clutch Control Valve
21— Engagement Override Valve
22— Lube Relief Valve
23— Port to PTO Lube
24— Clutch Lube Reduction Valve
25— Lube Valve Stop Pin
26— System Pressure Oil
27— Lube Pressure Oil
28— Pressure-Free Oil

Other valves are capable of "modulating" oil pressure for smooth engagement. This can be as simple as restricting oil flow with an orifice and cushioning element pressure with a spring accumulator.

Fig. 25 shows a type of modulating valve that has been used for decades, especially for traction clutch engagement. When the clutch valve is pushed to the right, it will allow oil from the pressure regulating valve to flow up to a passage to the clutch.

Continued on next page

POWER SHIFT TRANSMISSIONS

But we don't push the valve directly. Instead, we push against a spring inside the valve. The opposite end of the valve is exposed to the same pressure as the clutch. When pressure begins to build in the clutch, it pushes the valve to the left by compressing the spring. This shuts off flow to the clutch, so pressure will not go any higher until we push harder. The more we compress the spring, the higher it builds clutch pressure. When we let the pedal all the way up, it pushes the valve all the way to the right for maximum pressure. This arrangement has enabled two generations of operators to feather the pedal for smooth clutch engagement.

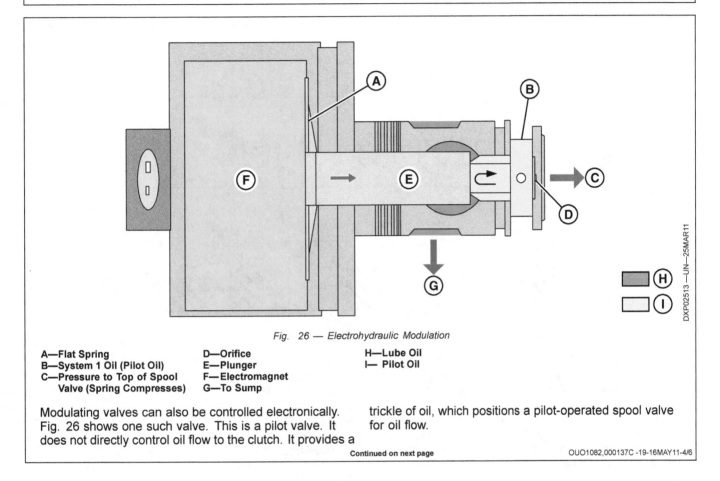

Fig. 26 — Electrohydraulic Modulation

A—Flat Spring
B—System 1 Oil (Pilot Oil)
C—Pressure to Top of Spool Valve (Spring Compresses)
D—Orifice
E—Plunger
F—Electromagnet
G—To Sump
H—Lube Oil
I—Pilot Oil

Modulating valves can also be controlled electronically. Fig. 26 shows one such valve. This is a pilot valve. It does not directly control oil flow to the clutch. It provides a trickle of oil, which positions a pilot-operated spool valve for oil flow.

Continued on next page

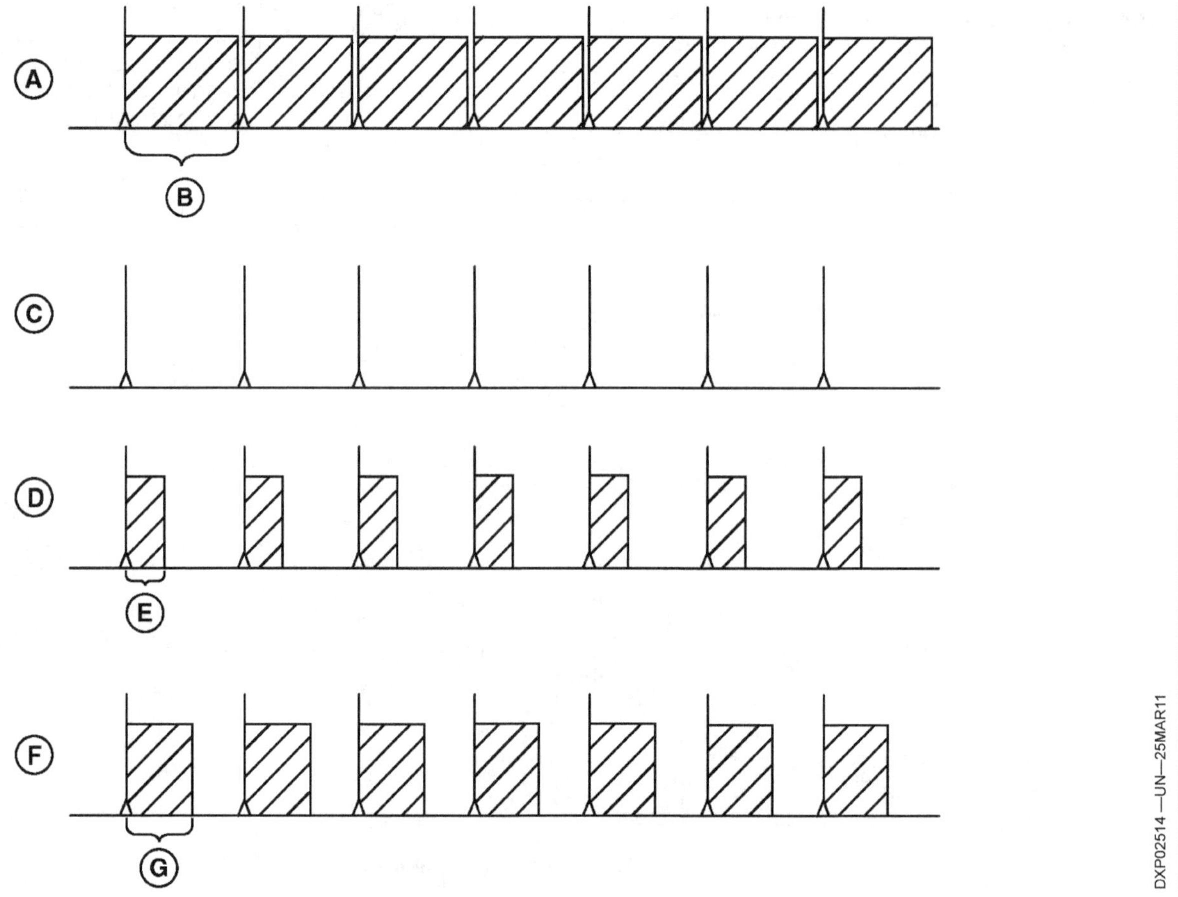

POWER SHIFT TRANSMISSIONS

Fig. 27 — Pulse Width Modulation

A—Signal for Full Pressure
B—Pulse Width at 100% Duty Cycle
C—Pulse Frequency
D—Signal for Low Pressure
E—Pulse Width at 36% Duty Cycle
F—Signal for Medium Pressure
G—Pulse Width at 57% Duty Cycle

Rather than varying voltage to the pilot valve, it uses pulse width modulation (Fig. 27). Electricity is turned on and off many times per second. The percentage of time it's turned on regulates valve position. The frequency is much too high for the valve to open and close each time, so it sort of floats in partial engagement. Spring force is balanced against electromagnetic force. Adjusting the percentage of time electricity is on, called duty cycle, regulates the valve very precisely.

Continued on next page

POWER SHIFT TRANSMISSIONS

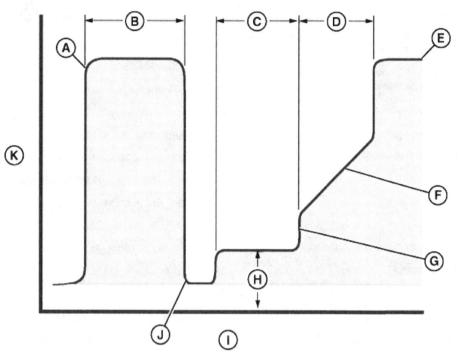

Fig. 28 — Modulating Valve Opening

A—Wide Open Valve to Get 90% Full (Calibration Time Less 10%)
B—Wake-Up Pulse
C—Final Fill
D—Modulation
E—Complete Engagement of Clutch
F—Adjusted Load Compensation and Variable Rate Modulation During Neutral to Gear Shifts
G—Clutch Exchange (Off-Going Clutch Is Sumped)
H—Calibration Pressure Plus 10% to Ensure Clutch Engagement
I—Time
J—Valve Setting
K—Command Pressure

Fig. 28 shows a typical valve command for shifting from one clutch to another:

- For the wake-up pulse, the valve is wide open. This pushes the clutch piston almost far enough to begin squeezing disks and plates, but not quite.
- Then a small flow of oil provides final fill. This clutch is ready to assume the load.
- As this clutch begins to engage, the clutch that was previously engaged is released.
- Engagement pressure is smoothly increased to a point where there should be no slippage, then opened wide.

CALIBRATION

Because of manufacturing tolerances and other variables, the transmission must be calibrated to establish a profile of normal performance. The wake-up pulse must precisely match fill volume, for instance. It must be recalibrated after any repairs, or shift quality might deteriorate.

Fortunately, calibration is usually easy. You just initiate the process and wait until it's done. Some machines automatically go into calibration mode anytime they're left running in neutral longer than a few minutes.

While it isn't necessary to know exactly what's happening during the calibration cycle, the logic is interesting. To calibrate the clutch valve in our previous example, the on-board computer might do something like this:

- Engage two other clutches that will lock the shaft so this clutch can't turn anything.
- Supply the valve a 1% duty cycle and wait a few seconds to see if engine speed drops.
- Increase the duty cycle to 2% and repeat.
- Continue increasing the duty cycle until engine speed drops 20 rpm.
- Measure fill volume by opening the valve and seeing how many milliseconds before engagement begins.

You can hear quirky little responses while the computer checks one element after another. When it's finished, it will know exactly how to regulate this transmission for optimum performance.

How much has changed since the days of simple gearboxes and shift levers!

POWER SHIFT TRANSMISSIONS

TROUBLESHOOTING

These troubleshooting charts are given as a general guide to the common failures of power shift transmissions.

Supplement these charts with specific troubleshooting from the machine technical manual.

MACHINE WON'T MOVE

1. Tow disconnect disengaged.
2. Cold weather starting clutch disengaged.
3. Failed clutch.
4. Park lock engaged.
5. No oil pressure.
6. Control linkage stuck or disconnected.
7. Blown packing or gaskets.
8. Oil filter plugged.

SHIFTS ERRATICALLY

1. Wrong setting of adjustable orifice.
2. Incorrect system oil pressure.
3. Failed or stuck accumulator.
4. Shift control disconnected or binding.

CLUTCHES OR BRAKES SLIPPING

1. Low oil pressure.
2. Defective gasket or seal.
3. Burned or warped clutch facings.
4. Excessive oil leakage at shift valves.

LOW SYSTEM PRESSURE

1. Plugged oil filter.
2. Low oil level.
3. Blown O-ring or gasket.
4. Plugged reservoir oil screen.
5. Air leak in pump suction line.
6. Wrong adjustment of pressure regulating valve.
7. Inching pedal valve leaking.
8. Stuck filter bypass valve.
9. Stuck cooler relief valve.
10. Failed or worn pump.
11. Pump not turning.

TRANSMISSION OVERHEATING

1. Reservoir oil level too low or too high.
2. Plugged oil filter.
3. Plugged core in oil cooler.
4. Not enough lubricating oil flow.
5. Faulty temperature gauge.
6. Clutch or brake dragging or slipping.
7. Cooler relief valve stuck open.

TEST YOURSELF

QUESTIONS

1. (Fill in the blanks.) In power shift transmissions, _____ _____ control the power flow, while the _____ _____ transmits the power flow.
2. What are the two basic types of gear sets used in power shift transmissions?
3. What do "underdrive" and "overdrive" mean?
4. Name the three main components of a planetary gear set.
5. (True or False?) When any two parts of a planetary are locked together, all three parts rotate as a unit.
6. What is the gear ratio of input to output when a planetary has two parts locked together?
7. What two jobs does oil do in a power shift transmission?

HYDROSTATIC DRIVES

INTRODUCTION

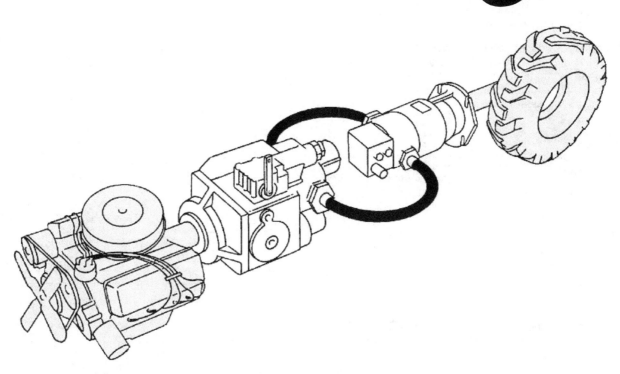

A hydrostatic drive uses oil pressure and flow to transmit power to the drive wheels of the machine. A hydraulic system — pump, motor, valves, oil lines — takes the place of the clutch and transmission.

The figure above illustrates the principles of a hydrostatic drive. Mechanical power from the engine is converted to hydraulic power by a pump. A motor then converts the hydraulic power back into mechanical power for the drive wheels.

Controls on the pump regulate the volume and direction of oil flow, thus controlling the speed and direction of machine travel. Speed is infinitely variable.

ADVANTAGES

The benefits are obvious:

- Convenience: Single lever control for stepless, clutchless, on-the-go changes of speed and direction.
- Infinite variability: From the slowest creeper speeds to transport, and everything in between.
- Flexible location: Those oil lines can be as long as required. They can go around corners in any direction.

They can connect assemblies that move relative to each other.

There are trade-offs, of course. Cost is higher than that of a simple gearbox. Efficiency is lower, especially under heavy load. Speed range is limited, sometimes requiring a gearbox in addition to the hydrostatic drive. Repairs require specialized workshops.

HYDROSTATIC DRIVES

HOW IT WORKS

Fig. 1 shows two basic types of hydraulic transmissions. In a hydrodynamic transmission, oil is directed at high speed against a turbine. This spins the turbine and turns the wheels. A hydrodynamic transmission is a non-positive drive, meaning output speed is flexible. Increasing load on the drive wheels can slow or even stop the turbine. Torque converters, covered in chapter 6, are hydrodynamic transmissions.

A hydrostatic transmission is a positive drive. The oil is trapped. Power is transmitted by oil pressure and flow, rather than by momentum from high speed. Except for slight internal leakage, output speed is controlled as precisely as in a gear drive.

A—Hose
B—Wheel
C—Hydrodynamic = Low Pressure and High Velocity
D—Engine
E—Pump
F—Motor
G—Wheel
H—Hydrostatic = High Pressure and Low Velocity

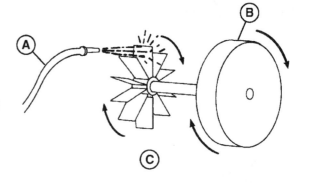
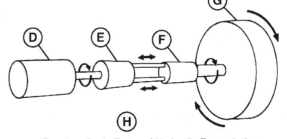

Fig. 1 — Basic Types of Hydraulic Transmission

BASICS

Liquid assumes the shape of its container (Fig. 2). Oil under pressure will flow in any direction into a passage of any size and shape.

Fig. 2 — Liquids Have No Shape of Their Own

For all practical purposes, liquid is incompressible. If you tried to reduce the size of a container full of oil, it would distort or burst the container (Fig. 3).

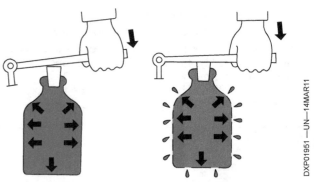

Fig. 3 — Liquids Will Not Compress

Continued on next page

Fig. 4 shows two cylinders connected by an oil line. If we push one piston in, oil pushes the other piston out. In its simplest form, this is the operating principle of a hydrostatic drive. The pump and motor both contain multiple pistons. They convert rotation into reciprocating motion and back again. But the bottom line is that a piston here pushes a piston there.

"Equal force out" is true only if the cylinders have the same diameter, which they typically have.

DXP02520 —UN—28MAR11

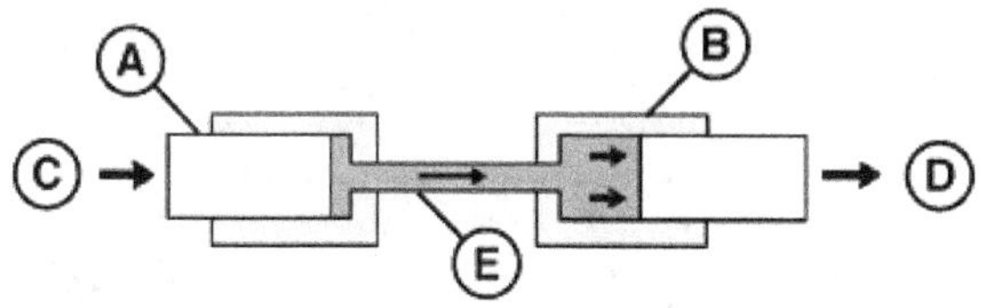

Fig. 4 — Two Cylinders Connected

A—Piston
B—Cylinder
C—Force In Here
D—Equal Force Out Here
E—Connecting Line

OUO1082,0001383 -19-16MAY11-3/8

Most hydrostatic pumps and motors are axial piston design, meaning the pistons move back and forth parallel to the shaft. Most of them have a swashplate at an angle on one end. The pistons are against the swashplate. As the pump or motor rotates, the pistons move back and forth.

Fig. 5 shows one pump piston and one motor piston. Remember, an oil passage is the only connection between the two. Step-by-step, let's watch what happens.

1. We push the pump piston downward. This is done by the engine turning the pump shaft.
2. The swashplate isn't moving. As the piston moves down, it is pushed to the right by the angled surface of the swashplate.
3. As the pump piston moves in, it forces oil out of the cylinder.
4. Oil moves through the connecting line from the pump to the motor.
5. Incoming oil pushes the motor piston to the right.

DXP02521 —UN—28MAR11

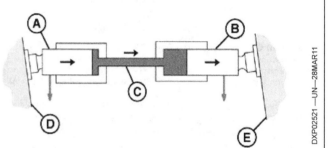

Fig. 5 — Two Connected Cylinders with Swashplates

A—Pump Piston
B—Motor Piston
C—Connecting Line
D—Pump Swashplate
E—Motor Swashplate

6. The only way the motor piston can move outward is by sliding down the angled surface of the swashplate, so the motor cylinder moves downward. This turns the motor shaft and drives the wheels.

Continued on next page

OUO1082,0001383 -19-16MAY11-4/8

HYDROSTATIC DRIVES

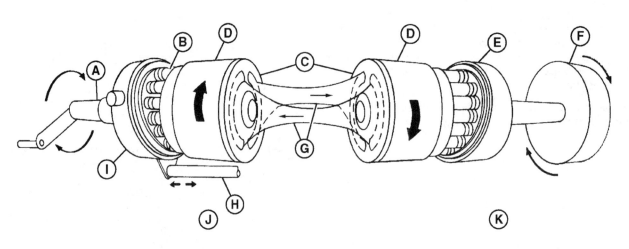

Fig. 6 — Variable-Displacement Pump Driving Fixed-Displacement Motor

A—Input Drive Shaft
B—Pistons
C—Valve Plates
D—Cylinder Block
E—Fixed Swashplate
F—Output Wheel
G—Oil Circuit
H—Swashplate Angle Control
I—Variable Swashplate
J—Pump (Variable Displacement)
K—Motor (Fixed Displacement)

Fig. 6 shows a pump and motor. Both have multiple pistons. The pistons are in a cylinder block, and the cylinder block is splined to the shaft. As the cylinder block rotates, the pistons slide in and out, controlled by the angle of the swashplate. Nothing else rotates. The swashplate may or may not have an adjustable angle, but it does not rotate.

The valve plate is held tightly against the cylinder block, but it does not rotate. A long opening on each side of the valve plate connects all the piston bores on that side.

With the swashplates angled as shown, pistons are fully extended at the top and fully retracted at the bottom.

With the shaft rotating as shown, pump pistons are pushed in by the swashplate as they move down on the far side. This forces oil out the slotted hole on the far side of the valve plate. Return oil from the motor flows into the near side of the pump valve plate, and pistons are extended as they move up the near side of the swashplate.

Let's look at the other half of the system. Oil is forced into the slot on the far side of the motor valve plate. This forces the pistons to extend. The only way they can move is by sliding upward on the angled swashplate. As pistons slide downward on the near side of the motor, the swashplate pushes them back in and returns oil to the pump inlet. In this example, the motor turns in the opposite direction from the pump.

Pump or motor capacity is expressed as "displacement." This is the volume of oil flow in one complete revolution. It is measured the same way as engine displacement:

Displacement = (piston area) x (length of stroke) x (number of pistons)

Continued on next page

HYDROSTATIC DRIVES

If the swashplate is non-adjustable, like the motor in Fig. 7, it is a fixed-displacement version.

If the swashplate angle is adjustable, like the pump, it is a variable-displacement version.

In almost all instances, the pump is variable-displacement and the motor is fixed-displacement.

When swashplate angle is reduced, piston stroke is shortened and flow is reduced. The machine slows down. When the swashplate is exactly vertical, flow is zero. The machine doesn't move.

If the swashplate angle moves beyond neutral to slope the other way, oil flows in the opposite direction. The machine moves in reverse.

Fig. 7 shows a typical pump for a hydrostatic drive. It's a variable-displacement pump with nine pistons. Swashplate angle is adjusted by a servo piston.

Unlike the previous example, the swashplate pivot axis is vertical instead of horizontal. Therefore, inlet and outlet slots in the valve plate are now at top and bottom instead of on the sides.

Port A and Port B are oil passages to the motor.

The charging pump is part of the external system, to be covered a little later.

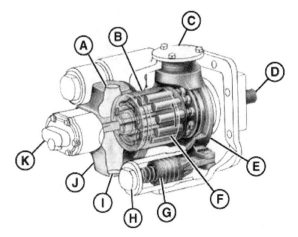

Fig. 7 — Variable-Displacement Pump – Axial Piston Type

A—Port A
B—Cylinder Block
C—Pivot
D—Drive Shaft
E—Swashplate
F—Piston
G—Servo Piston
H—Servo Cylinder
I— Port B
J—End Cap
K—Charging Pump

Fig. 8 shows a typical motor for a hydrostatic drive. It's a fixed-displacement motor with nine pistons. Swashplate angle is not adjustable.

Port A and Port B are oil passages to the pump.

The high-pressure relief valve prevents damage if the load is so great that the wheels can't turn. Operating pressure in a hydrostatic drive is extremely high compared to other systems. Relief valve pressure may be 7000–8000 psi (480–550 bar).

The charge pressure control valve and shuttle valve are part of the external system.

A—Manifold
B—Port A
C—End Cap
D—Cylinder Block
E—Fixed Swashplate
F—Pistons
G—Valve Plate
H—Port B
I— Charge Pressure Control Valve
J— Shuttle Valve
K—High-Pressure Relief Valve

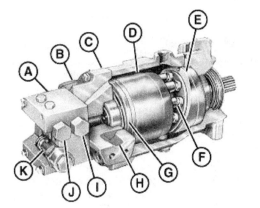

Fig. 8 — Fixed-Displacement Motor – Axial Piston Type

Continued on next page

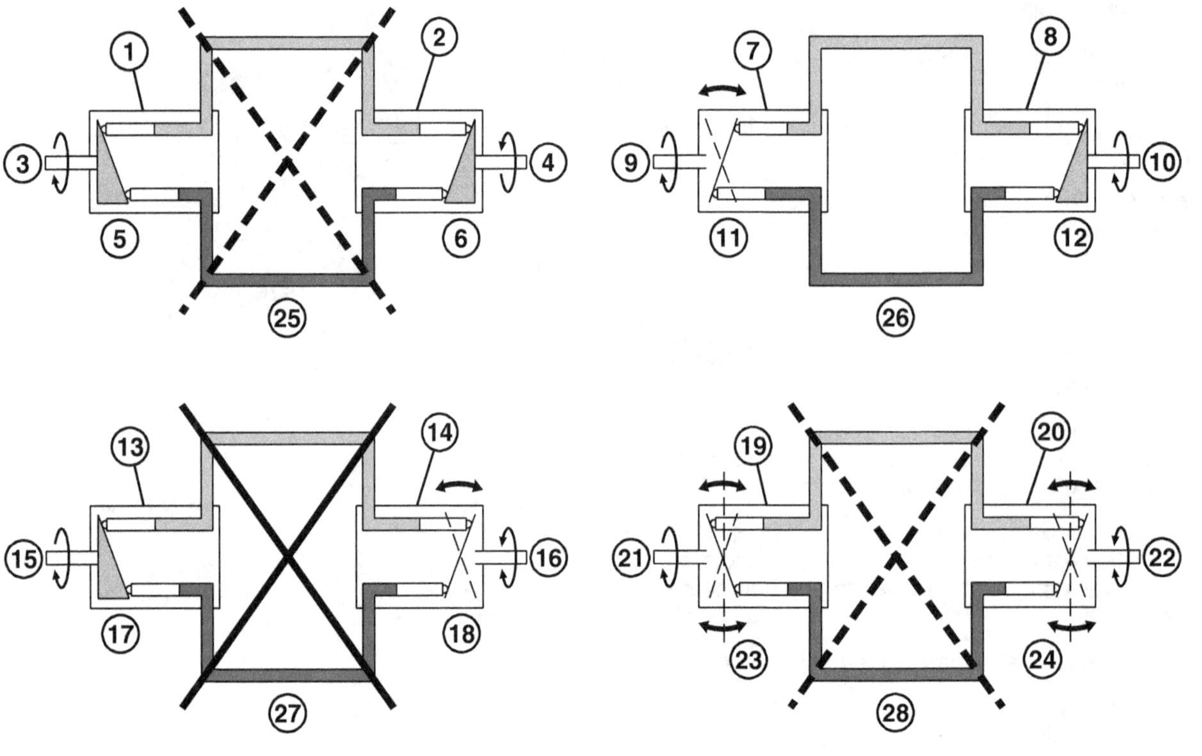

Fig. 9 — Pump-Motor Combinations for Hydrostatic Drives

1— Pump
2— Motor
3— Input
4— Output
5— Swashplate Angle Always the Same
6— Swashplate Angle Always the Same
7— Pump
8— Motor
9— Input
10— Output
11— Swashplate Angle Can Be Changed
12— Swashplate Angle Always the Same
13— Pump
14— Motor
15— Input
16— Output
17— Swashplate Angle Always the Same
18— Swashplate Angle Can Be Changed
19— Pump
20— Motor
21— Input
22— Output
23— Swashplate Angle Can Be Changed
24— Swashplate Angle Can Be Changed
25— Fixed-Displacement Pump, Fixed-Displacement Motor
26— Variable-Displacement Pump, Fixed-Displacement Motor
27— Fixed-Displacement Pump, Variable-Displacement Motor
28— Variable-Displacement Pump, Variable-Displacement Motor

Fig. 9 illustrates four pump-motor combinations that would theoretically be possible. The great majority of hydrostatic drives are a Variable-Displacement Pump, Fixed-Displacement Motor (26) combination. A variable-displacement pump controls the direction and speed of a fixed-displacement motor.

If both pump and motor were fixed-displacement (25), you couldn't change speed or direction. This is a convenient way to power a mechanism where shafts or chains wouldn't be practical, but it can't replace a transmission. This approach is used on center pivot mowers, connecting the PTO to a machine that swings from side to side. It can power a boom-mounted cutter or front-mounted snowblower.

A Fixed-Displacement Pump, Variable-Displacement Motor (27) is unworkable. Swashplate angle in a motor must never approach zero. Oil from the pump would have no place to go, causing system failure.

Also a Variable-Displacement Pump, Variable-Displacement Motor (28) wouldn't work as shown, because the motor swashplate angle must never approach zero. However, the angle can be varied within a narrow band to extend the speed range.

HYDROSTATIC DRIVES

COMPLETE SYSTEM

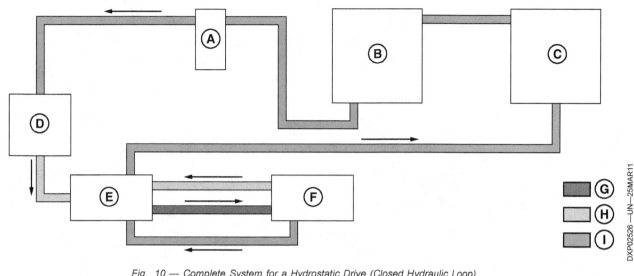

Fig. 10 — Complete System for a Hydrostatic Drive (Closed Hydraulic Loop)

A—Filter
B—Reservoir
C—Cooler
D—Charge Pump
E—Pump
F—Motor
G—High Pressure Oil
H—Low Pressure Oil
I— Pressure Free Oil

Let's back up for a wider view. Fig. 10 shows how the hydrostatic pump and motor might fit into a larger system.

The pump and motor are almost a closed loop by themselves. Oil from the pump flows through one line to the motor and back through the other line. Most of the oil remains in this circuit.

However, there is a continuous exchange between the hydrostatic loop and the rest of the system. This is done to cool and filter the oil.

The charge pump, filter, cooler, and reservoir are often shared with other systems. For instance, the charge pump might supply oil to a hydraulic pump in addition to the hydrostatic pump. The other pump is for power steering, hydraulic cylinders, etc.

HYDROSTATIC DRIVES

OPERATION

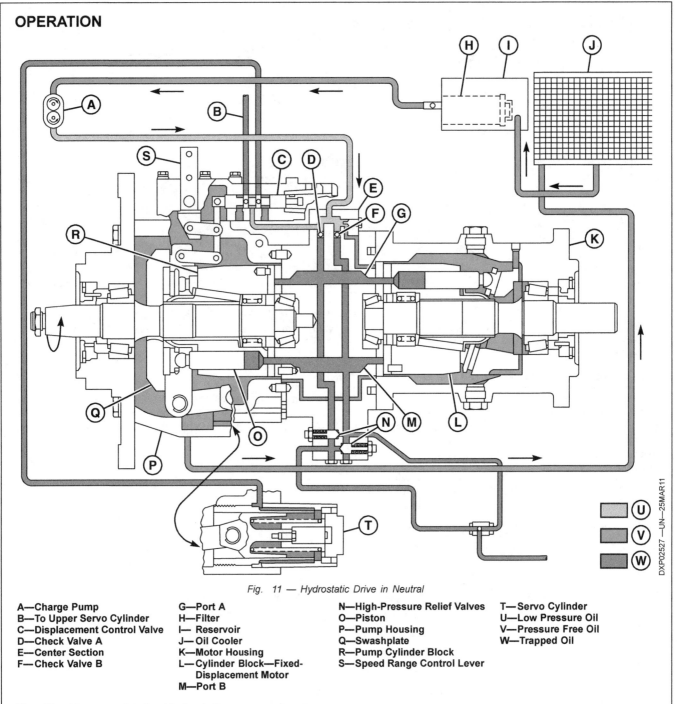

Fig. 11 — Hydrostatic Drive in Neutral

A—Charge Pump
B—To Upper Servo Cylinder
C—Displacement Control Valve
D—Check Valve A
E—Center Section
F—Check Valve B
G—Port A
H—Filter
I—Reservoir
J—Oil Cooler
K—Motor Housing
L—Cylinder Block—Fixed-Displacement Motor
M—Port B
N—High-Pressure Relief Valves
O—Piston
P—Pump Housing
Q—Swashplate
R—Pump Cylinder Block
S—Speed Range Control Lever
T—Servo Cylinder
U—Low Pressure Oil
V—Pressure Free Oil
W—Trapped Oil

Fig. 11 adds more details. Hydrostatic pump and motor are built into one assembly, with only an oil manifold between them. This is the common arrangement, unless drive wheels are far removed from the engine.

The diagram includes the complete system — charge pump, filter, etc. It also includes the servo cylinder which adjusts swashplate angle, the displacement control which regulates the servo cylinder, and the speed control lever which the operator uses to select travel speed.

The speed control lever is in neutral, meaning the operator does not want the machine to move. The swashplate is exactly vertical. The pump cylinder block is rotating, but it is not pumping any oil. The motor is not turning, because there is no flow from the pump.

Green indicates "trapped" oil. Until something changes, everything between pump pistons and motor pistons will be trapped oil.

Continued on next page

HYDROSTATIC DRIVES

Pink indicates low pressure charge oil. There's a constant flow from the charge pump through the hydrostatic pump body and back through the cooler to the reservoir. It is available if needed to operate the servo cylinder. Since none is needed, it flows through the displacement control valve into the pump body. It also lubricates bearings in the pump and motor.

HYDROSTATIC DRIVES

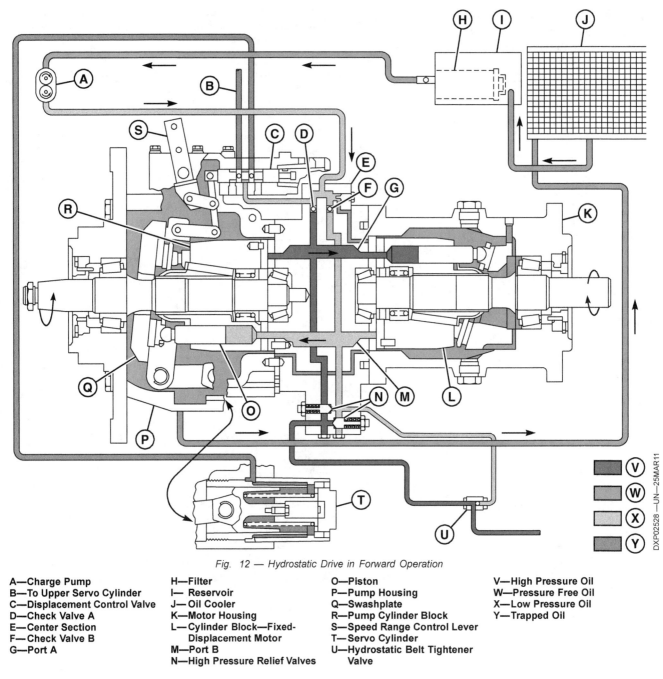

Fig. 12 — Hydrostatic Drive in Forward Operation

A—Charge Pump
B—To Upper Servo Cylinder
C—Displacement Control Valve
D—Check Valve A
E—Center Section
F—Check Valve B
G—Port A
H—Filter
I—Reservoir
J—Oil Cooler
K—Motor Housing
L—Cylinder Block—Fixed-Displacement Motor
M—Port B
N—High Pressure Relief Valves
O—Piston
P—Pump Housing
Q—Swashplate
R—Pump Cylinder Block
S—Speed Range Control Lever
T—Servo Cylinder
U—Hydrostatic Belt Tightener Valve
V—High Pressure Oil
W—Pressure Free Oil
X—Low Pressure Oil
Y—Trapped Oil

Fig. 12 shows what happens when the operator pushes the control lever for forward travel.

Actually, two things have already happened:

- The speed control lever pushed the displacement control valve rearward. This forced charge oil out to the lower servo cylinder and allowed oil to return from the upper servo cylinder.
- The servo cylinders tilted the swashplate back at the top. Movement of the swashplate pulled the displacement control valve forward to its original position.

So the displacement control valve is back to neutral. The swashplate is at a new angle. The servo cylinders are in new positions, once again held by trapped oil until the speed control lever moves again.

Pump rotation is counterclockwise as viewed from the front. With the swashplate tilted back at the top, pistons moving up the near side are forced inward. This pumps oil out the slot on the near side of the valve plate, port A.

Continued on next page

Oil flows straight through the center manifold to the slot on the near side of the motor valve plate, pushing the pistons on the near side of the motor cylinder block. The motor swashplate is always tilted back at the top, so the motor rotates in the same direction as the pump.

Return oil is pushed back from the far side of the motor through port B to the far side of the pump. If any make-up oil is needed due to leakage, charge oil can unseat check valve B and flow into the return circuit.

If extreme loads push pressure too high, oil in port A can unseat the high-pressure relief valve and dump oil into port B.

Continued on next page

HYDROSTATIC DRIVES

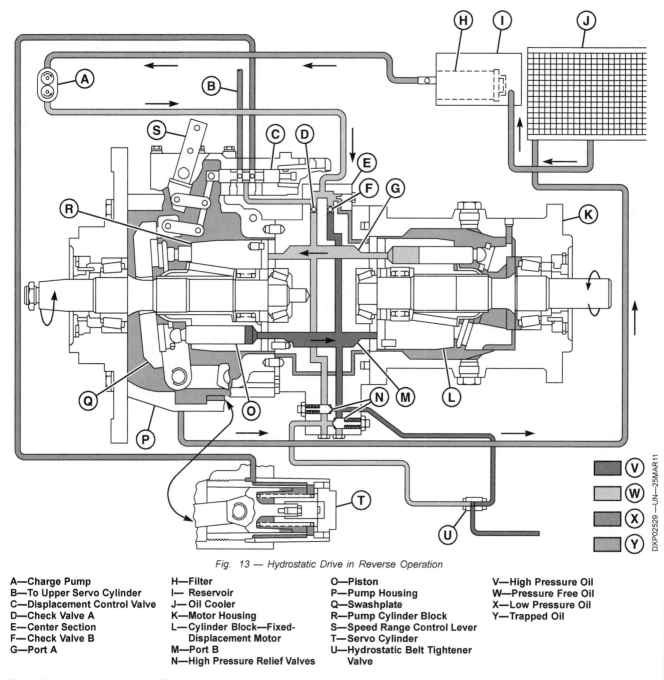

Fig. 13 — Hydrostatic Drive in Reverse Operation

A—Charge Pump
B—To Upper Servo Cylinder
C—Displacement Control Valve
D—Check Valve A
E—Center Section
F—Check Valve B
G—Port A
H—Filter
I—Reservoir
J—Oil Cooler
K—Motor Housing
L—Cylinder Block—Fixed-Displacement Motor
M—Port B
N—High Pressure Relief Valves
O—Piston
P—Pump Housing
Q—Swashplate
R—Pump Cylinder Block
S—Speed Range Control Lever
T—Servo Cylinder
U—Hydrostatic Belt Tightener Valve
V—High Pressure Oil
W—Pressure Free Oil
X—Low Pressure Oil
Y—Trapped Oil

Fig. 13 shows the reverse. The operator has pulled the speed control lever back past neutral. This pulled the displacement control valve forward, which forced charge oil out to the upper servo cylinder and allowed oil to return from the lower servo cylinder. The swashplate tilted forward at the top, which pushed the displacement control valve back to its neutral position.

The pump is still rotating counterclockwise, but with the swashplate tilted forward, it now pumps oil out on the far side to port B.

Pressure oil on the far side of the motor pushes pistons out. The motor swashplate angle remains the same, so now the motor rotates in the opposite direction from the pump. Everything else remains the same.

In summary, changing the direction the pump swashplate is tilted will change the direction the vehicle moves. Changing the degree it is tilted will change the speed.

OUO1082,0001385 -19-16MAY11-5/5

5-12

HYDROSTATIC DRIVE AXLES

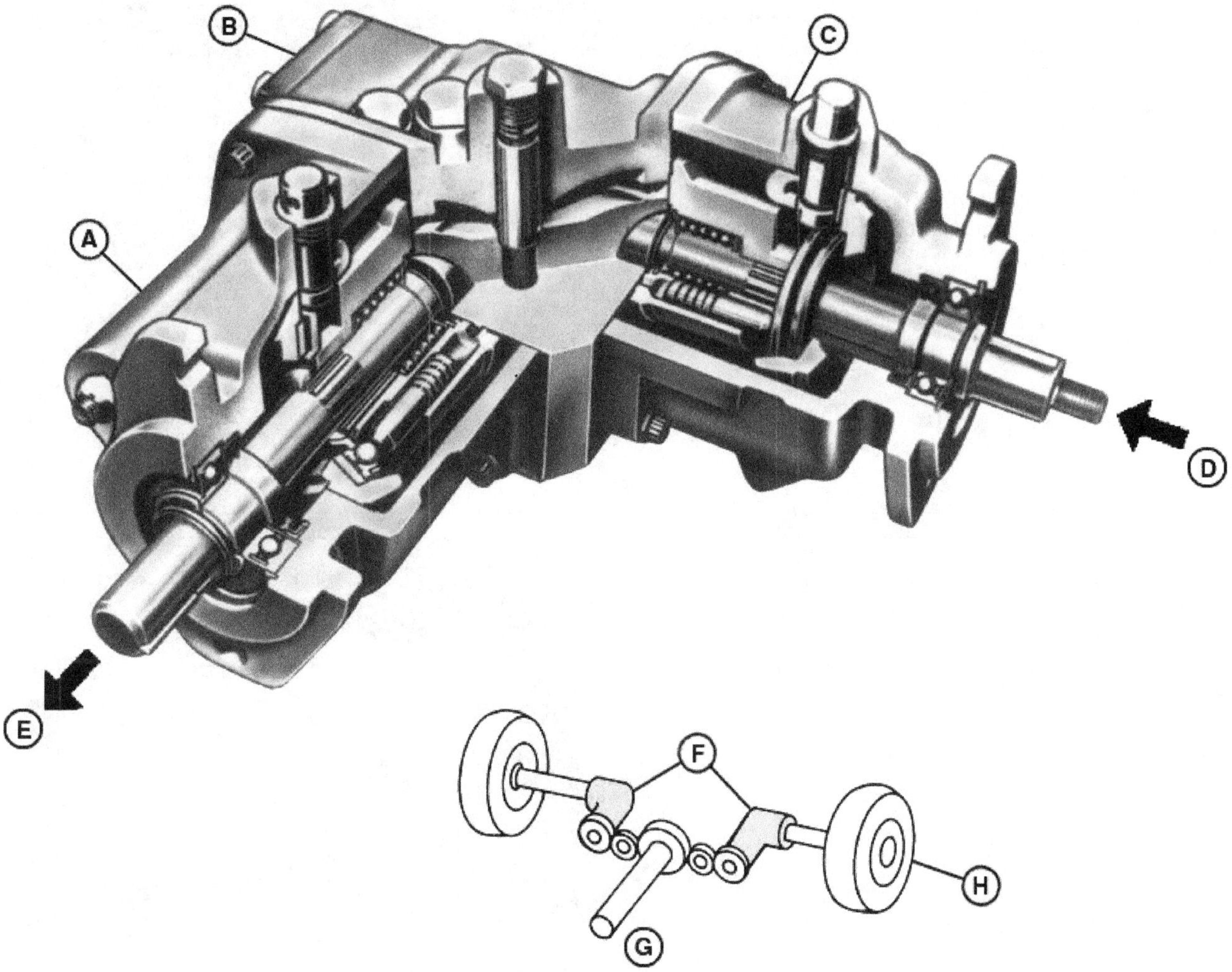

Fig. 14 — Integral Pump-Motor for Hydrostatic Drive Axle

A—Axial Piston Motor
B—Charge Pump
C—Axial Piston Pump
D—Input from Engine
E—Output to Drive Axle
F—Two Hydrostatic Drives
G—From Engine
H—Drive Wheel

Sometimes a machine uses two hydrostatic drives, one for each side. This provides steering in addition to power transmission, and it allows zero turning radius — pivoting in one spot. Examples include dozers, skid-steer loaders, windrowers, and certain lawn mowers.

Fig. 14 illustrates one example of hydrostatic drive axles. Pump and motor are built into one assembly, but they are at right angles. There is one assembly on each side of the vehicle, and the engine drives both pumps.

Both pump and motor have swashplates. The pump swashplate controls direction and speed.

This is a rare instance where the motor swashplate angle is also adjustable. Not reversible, but adjustable within a narrow band to extend the speed range of the vehicle.

Most such vehicles use two levers for total control — direction, speed, and steering. One lever is attached to each hydrostatic drive unit. Sometimes other functions such as loader boom and bucket cylinders are controlled by tilting the same two levers side-to-side. Just fasten your seat belt, brace your feet, and have all controls in the palms of your hands.

HYDROSTATIC DRIVES

OTHER TYPES OF PUMPS AND MOTORS

Swashplates are by far the most common design for hydrostatic pumps and motors, but they are not universal. A bent-axis design offers higher efficiency and wider operating range.

Fig. 15 is a cutaway view of a bent-axis motor. The cylinder block, identified here as "rotating group," is at a 45° angle to the shaft it turns. The bent axis does exactly the same job as a swashplate, but without the friction loss caused by sliding parts under extremely high forces.

A—Fixed Unit Gear
B—Rotating Group
C—Ring Gear

Fig. 15 — Bent-Axis Hydrostatic Motor

Fig. 16 is a cutaway view of a bent-axis pump. It's a variable-displacement pump. The cylinder block, or rotating group, is in a large yoke that pivots side-to-side. It can swing 45° in either direction, a much wider range than is possible with swashplates.

A—Yoke
B—Rotating Group

Fig. 16 — Bent-Axis Hydrostatic Pump

Continued on next page

HYDROSTATIC DRIVES

This particular pump is controlled electrohydraulically. Fig. 17 shows the pilot-operated spool valve and pistons that swing the yoke.

A—Hydro Control Valve
B—Bias Piston
C—Spring
D—Follower
E—Cam
F—Yoke
G—Servo Piston
H—System 1 Oil
I—From Hydro Control Pilot Valve
J—Sump

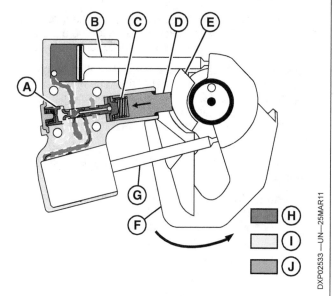

Fig. 17 — Yoke Angle Controls

The complete system is shown in Fig. 18 — pump, motor, and controls in one assembly. The pump yoke is at lower left in the photo.

Even a gear motor could be used in a hydrostatic drive, but internal leakage prevents the precision most drives require.

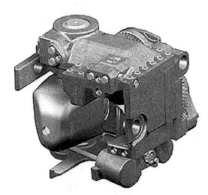

Fig. 18 — Bent-Axis Pump and Motor Assembly

Continued on next page

HYDROSTATIC DRIVES

CAM LOBE MOTOR

Another unusual approach uses the cam lobe motor (Figs. 19 and 20). The motor shaft is the axle for the drive wheel. Both steerable and non-steerable designs are available.

A carrier with 12 pistons rotates inside a circular cam with 15 lobes. The cam is stationary. The piston carrier is splined to the axle. Pressure oil forces the pistons outward when they are rolling "down" off the cam lobes. Pistons push return oil out when they are rolling "up" onto the cam lobes.

With 12 pistons, you can see that 3 are in each of these situations:

- 3 pistons (A) are on the power stroke.
- 3 pistons (B) are fully extended.
- 3 pistons (C) are on the return stroke.
- 3 pistons (D) are fully retracted.

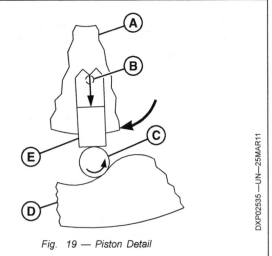

Fig. 19 — Piston Detail

A—Carrier
B—Manifold
C—Piston Follower
D—Cam (Fixed)
E—Piston

Continued on next page

HYDROSTATIC DRIVES

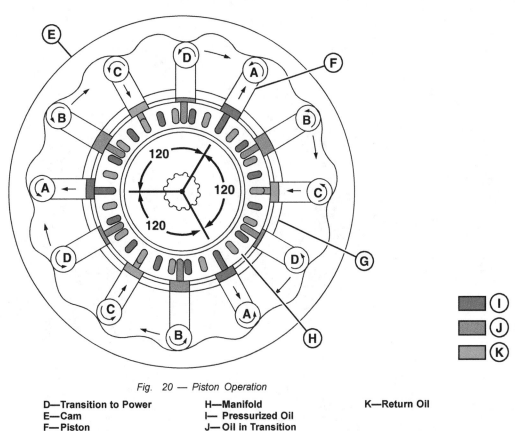

Fig. 20 — Piston Operation

A—Power Stroke
B—Transition to Return
C—Return Stroke
D—Transition to Power
E—Cam
F—Piston
G—Carrier
H—Manifold
I—Pressurized Oil
J—Oil in Transition
K—Return Oil

The piston carrier rotates alongside a stationary oil manifold. The manifold has 15 pressure ports and 15 return ports. Each piston aligns with a pressure port when it's rolling down off a cam lobe and with a return port when it's rolling up onto a cam lobe.

There is an overlap to ensure smooth, continuous power. By the time three pistons complete the power stroke, three more have begun.

Of course, pressure and return oil trade places and reverse flow direction to back up.

Cam lobe motors are often used only as a "booster" such as rear-wheel-assist on a combine. The large front wheels control vehicle speed. The rear-wheel-assist can be turned on and off as needed for traction conditions. In this case, pressure oil for the cam lobe motor comes from the hydraulic system, not from a hydrostatic pump. It produces enough torque to assist the drive axle, but not enough to spin the wheels.

Continued on next page

HYDROSTATIC DRIVES

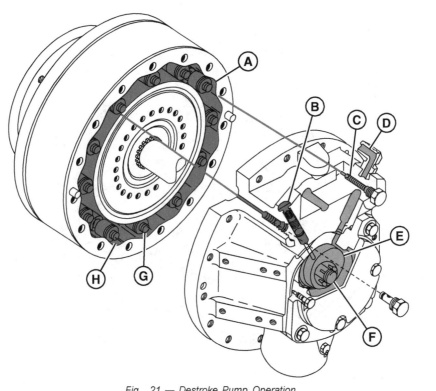

Fig. 21 — Destroke Pump Operation

A—Outer Case
B—Destroke Pump
C—Outlet Check Valve
D—Drain Line
E—Eccentric Washer
F—Axle
G—Inlet Check Valve
H—Cam Lobe Piston
I— Destroke-Pressure Oil
J—Low Pressure Oil

If a cam lobe motor is switched off, there is generally a provision to hold the pistons away from the cam lobes. This can be done by surrounding the piston carrier with enough pressure to push the pistons in. Fig. 21 illustrates a destroke mechanism. An eccentric washer on the axle operates the destroke pump plunger to pressurize the outer case. The outlet check valve regulates pressure.

Continued on next page

HYDROSTATIC DRIVES

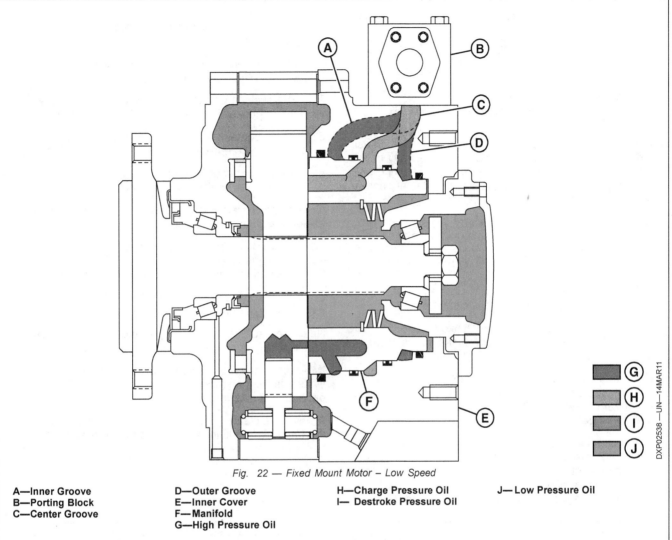

Fig. 22 — Fixed Mount Motor – Low Speed

A—Inner Groove
B—Porting Block
C—Center Groove
D—Outer Groove
E—Inner Cover
F—Manifold
G—High Pressure Oil
H—Charge Pressure Oil
I—Destroke Pressure Oil
J—Low Pressure Oil

Fig. 22 is a cross-section view of a cam lobe motor. It illustrates the optional three-speed porting block to expand the operating range. Ten manifold pressure ports are connected to the outer groove and five to the inner groove.

Pressure oil can be delivered to all 15 ports for highest torque and lowest speed, or only 10, or only 5. This results in three pistons on power stroke at any time, or only two, or only one. The fastest speed is triple the slowest, but torque is only one-third as much.

HYDROSTATIC DRIVES

MAINTENANCE OF HYDROSTATIC DRIVES

Any hydraulic system is fairly easy to maintain: the fluid provides a lubricant and protects against overload. But like any other mechanism, it must be operated properly.

You can damage a hydraulic system by too much **speed**, too much **heat**, too much **pressure**, or too much **contamination**.

Keep the hydrostatic system clean. Impurities such as dirt, lint, and chaff cause more damage than any one other thing. Seal any openings when doing service work to prevent dirt from entering the system. Be very careful that all parts are clean before replacing them.

NOTE: *Any internal service on either the pump or motor must be done under extremely clean conditions. The tolerances in these units are similar to those in diesel injection pumps, and equal care must be taken.*

Before removing any part of the hydrostatic system for servicing, be sure that the area is clean.

Use steam cleaning equipment, if it is available. However, do not let water enter the system. Make sure all hose and line connections are tight.

If steam cleaning is not possible, fuel oil or a suitable solvent may be used. Do not use paint thinner or acetone. Be certain to remove all loose dirt and foreign material that might get into the system when it is opened.

Use a clean work bench or table when disassembling the pump or motor for servicing. Do not perform internal service work on the shop floor or on the ground or where there is danger of dust or dirt being blown into parts. Be sure all tools are clean and free of grease and dirt.

Have these items available before disassembling the pump or motor:

1. Several clean plastic plugs to seal openings when removing hydraulic lines or hoses.

2. Several clean plastic bags to place over open ends of lines or hoses. Secure bags with rubber bands.

3. A clean container of solvent to wash internal parts. Use clean compressed air to dry parts after washing.

4. A clean container of transmission fluid to lubricate internal parts as they are assembled.

5. A clean container of petroleum jelly (petrolatum) to lubricate surfaces where noted during reassembly.

Whenever units are serviced, always install new O-rings, seals, and gaskets during reassembly. This will provide tight seals for mating parts and eliminate leakage.

Do not service the hydrostatic drive system with the machine engine operating unless specifically required.

Never operate the hydraulic system empty. Always check the oil supply after servicing the system.

If the machine is to be stored, be sure the hydraulic system is filled and the reservoir cap is tight. The oil will prevent rust and corrosion.

Always use clean oil in the system. Be careful to use the transmission fluid recommended in the machine operator's manual.

TESTING THE SYSTEM

The only testing required for many hydrostatic drives is for charging and operating oil pressures. Other systems may also require tests for oil flow rate and oil temperature.

Use the proper test equipment and procedures as given in the machine technical manual.

HYDROSTATIC DRIVES

SAFETY RULES

You must recognize potential hazards and take necessary action to avoid injury. Always observe the following safety precautions.

1. Know the pinch points on rotating parts (Fig. 23).

2. Always relieve the pressure in a hydraulic system before loosening, tightening, removing, or adjusting fittings and components.

A—Pinch Points
B—Belt Drive
C—Chain Drive
D—Gear Drive
E—Feed Roll

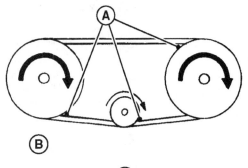
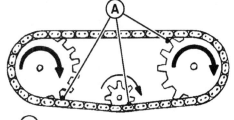
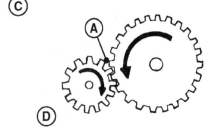
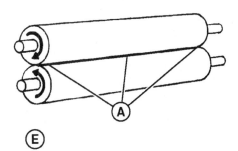

Fig. 23 — Pinch Points on Rotating Parts Can Catch Clothing, Hands, Arms, and Feet

Continued on next page

HYDROSTATIC DRIVES

3. Use caution when removing any device connected to a spring (Fig. 24).

4. Use proper tools to assist you in removing or replacing spring-loaded devices.

A—Pivot Point
B—Spring Under Tension
C—Spring Under Compression

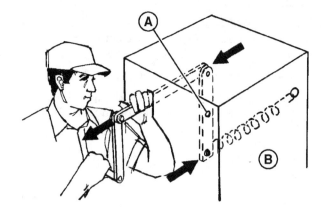

Fig. 24 — Be Sure You Know What Can Happen Before Disconnecting Any Part Attached to a Spring

HYDROSTATIC DRIVES

TROUBLESHOOTING OF HYDROSTATIC DRIVES

Follow this general guide for major problems and remedies. For specifics, see the machine technical manual.

NOTE: These troubleshooting charts include some items that will not apply to simpler drives.

HYDROSTATIC DRIVE TROUBLESHOOTING

CAUSE OR SYMPTOM	REMEDY
MACHINE WILL NOT MOVE IN EITHER DIRECTION	
System low on oil.	Fill with oil. Check for oil leaks.
Faulty control linkage.	Be sure couplings from engine to pump and from motor to gear train are installed correctly and not slipping or broken.
Low or zero charge pressure.	Check charge pressure control valve. Also look for binding pump drive shaft or oil line restrictions.
Low or fluctuating charge pressure.	Check for air in system (if noisy). Also look for damaged charge pressure control valve (stuck open) or faulty check valves.
MACHINE MOVES IN ONE DIRECTION ONLY	
Faulty control linkage.	Be certain control linkage is connected properly and not binding.
High-pressure relief valve stuck open.	Replace valve that is stuck open.
One faulty check valve.	Replace both check valves.
Faulty displacement control valve.	Repair displacement control valve.
NEUTRAL HARD TO FIND	
Faulty speed control linkage.	Adjust control linkage. Be certain linkage is not binding. Also, be sure displacement control valve is adjusted properly. Check pump servo cylinders for proper adjustment.
SYSTEM OPERATING HOT	
Oil level low.	Fill with oil.
Cooler clogged.	Clean cooler air passages.
Engine fan belt slipping or broken.	Tighten or replace fan belt.
Internal leaks (usually shown by loss of acceleration and power).	Check for stuck high-pressure relief valve. Replace relief valve. Check for internal damage in pump or motor.
SYSTEM NOISY	
Air in system.	Check oil supply. Be sure all air is out of system. Check for loose fittings or damaged lines and hoses.
HIGH LOSS OF OIL	
Loose connections or leaking lines and hoses.	Tighten connections. Replace damaged lines and hoses. Check O-rings and seats.

TEST YOURSELF

QUESTIONS

1. (Fill in each blank with "high" or "low.") Hydrostatic drives use fluids at _____ velocity but relatively _____ pressure.

2. How is a torque converter different from a hydrostatic drive as defined in Question No. 1?

3. What team of components is the heart of the hydrostatic drive?

4. Does a pump with a movable swashplate have a fixed or variable displacement?

5. What does the charge pump do for the hydrostatic drive oil circuit?

6. What are the two types of cam lobe motors?

7. How is the power kept constant to the axle on a cam lobe motor?

8. What is the function of a destroke pump in a cam lobe motor?

TORQUE CONVERTERS

INTRODUCTION

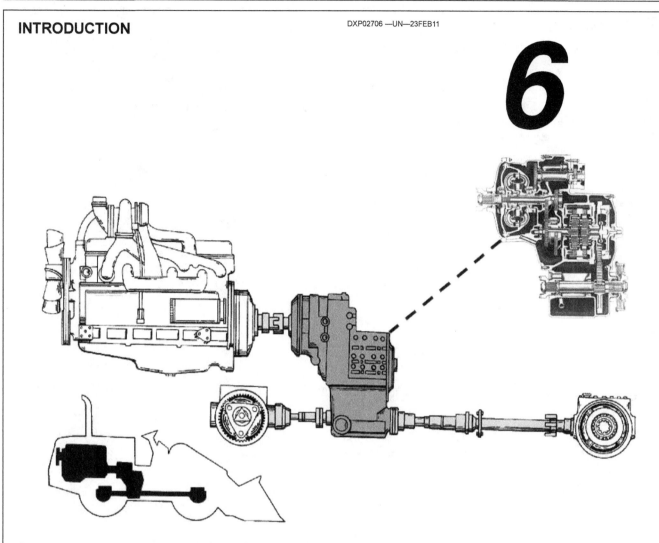

A torque converter is an automatic fluid drive. It transmits engine torque by means of hydraulic force, shifting smoothly through an infinite number of speeds.

The automatic transmission of an automobile automatically shifts gears in response to torque requirements in addition to the automatic response of the torque converter, which is part of the automobile's automatic transmission system.

Actually, a gear train is used with the torque converter to give extra speed ranges. But no gear train could give the infinite variations in speed and torque of a torque converter.

Acting as a clutch, the torque converter connects and disconnects power between the engine and the gear train. As a transmission, the converter gives many more speed ratios than are practical with a strictly mechanical gearbox.

To compare a torque converter with a hydrostatic drive (chapter 5), use this rule of thumb:

Hydrostatic drives are driven by fluids at *high pressure* but at relatively *low velocity*.

Torque converters are driven at *low pressure* but at *high velocity*.

Here are the formulas:

- Hydrostatic Drive = High Pressure + Low Velocity
- Torque Converter = Low Pressure + High Velocity

A torque converter is perhaps the ideal choice for jobs with momentary heavy loads, frequent stops, shuttle shifting, and driving place-to-place. Jobs an end loader might do.

But it isn't ideal for every job. Continuous heavy resistance, such as pulling a plow, would lead to oil heating and poor fuel economy.

TORQUE CONVERTERS

HOW IT WORKS

To understand a torque converter, we first must look at a basic fluid coupling.

The basic principles of all fluid couplings are shown in Fig. 1.

At the top, a fluid at high velocity strikes a turbine and forces it to turn, driving the wheel. Thus torque is transmitted by a fluid.

To change this torque, the velocity of the fluid must change. At low velocity, the fluid will not even move the turbine. At higher velocity, the turbine starts turning and the wheel picks up speed.

A—Hose
B—Wheel
C—Fluids at High Velocities Can Transmit Power
D—One Part Can Drive Another by Force of Air—or Oil

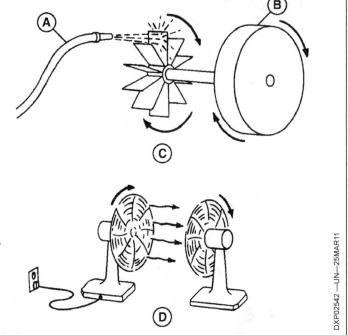

Fig. 1 — Basic Principles of a Fluid Coupling

Continued on next page

TORQUE CONVERTERS

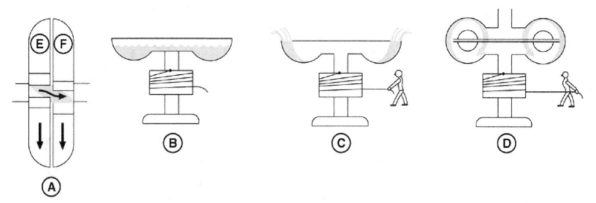

Fig. 2 — Operation of a Fluid Coupling

A—Fluid Coupling
B—Fluid Lies Level in Bowl
C—Bowl Is Spun and Fluid Spills Out
D—Torque Is Transmitted to Upper Bowl by Force of Fluid
E—Pump
F—Turbine

This principle is used in a fluid coupling as follows:

Inside an oil-filled housing (A, Fig. 2) are two parts: the driving half, or pump (impeller), and the driven half, or turbine.

As the pump is turned by the engine, centrifugal force causes oil to be forced radially outward, crossing over and striking the vanes of the turbine. This rotates the turbine in the same direction and so couples the power.

Fig. 2, B–D explain how the flow of oil drives the turbine:

- In "B," fluid is placed in a bowl and lies level.
- In "C," the bowl is spun rapidly and centrifugal force causes the fluid to climb up and spill over the outside edge of the bowl.
- In "D," another bowl is placed down over the first one. Now when the bowls are spun, an axial flow or circuit is created and turning force is transmitted between the driving bowl and the driven bowl.

Torque is thus transmitted, but it is not increased.

This is where the torque converter goes beyond the basic fluid coupling, for the converter can multiply torque.

A torque converter (Fig. 3) looks much like the fluid coupling we have just described. The main difference is that the torque converter has — in addition to the driving pump and the driven turbine — a set of blades or vanes called a stator.

The stator vanes change the direction of oil flow after it has gone through the turbine and sends it back to the pump. This enables the pump to increase the twisting force or multiply the torque.

Since the converter is a closed unit, this flow is repeated continuously. Many streams of fluid act against many vanes at once and this is what gives the power to drive a heavy machine.

A—Turbine
B—Pump (Impeller)
C—To Drive Wheels
D—Stator
E—Pump
F—Turbine
G—From Engine
H—Stator

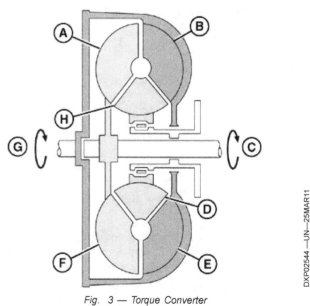

Fig. 3 — Torque Converter

Continued on next page

TORQUE CONVERTERS

INCREASING THE TORQUE

Remember that the pump is driven by the engine, while the turbine receives fluid energy from the pump and sends it to the drive wheels.

Also remember how centrifugal force sets up a continuous circular flow in the coupling (Fig. 4).

This circular flow of oil between the pump and turbine is called vortex flow.

Another flow is set up around the pump and turbine to form a coupling; this is called rotary flow.

The action of these combined oil flows will transmit torque *but not increase it.*

Increasing the torque is where the stator comes into play.

A—Rotary Oil Flow
B—Driving Member Pump (Impeller)
C—Vortex Oil Flow

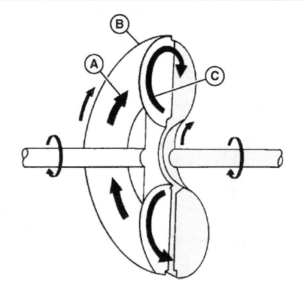

Fig. 4 — Vortex and Rotary Flow in a Fluid Coupling

OIL FLOW IN CONVERTER

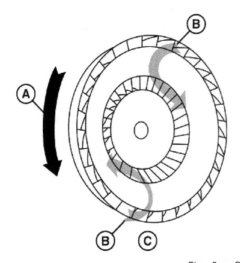 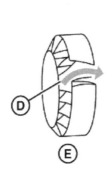 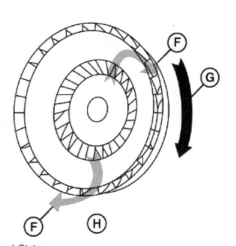

Fig. 5 — Oil Flow Through Pump, Turbine, and Stator

A—Rotation
B—Entrance
C—Turbine
D—Oil from Turbine
E—Stator (Stationary)
F—Exit
G—Rotation
H—Pump

Let's look at the flow of oil in the converter during two cycles:

Fig. 5 shows how oil emerges from the turbine in reverse compared to pump rotation. Unless this oil flow is turned around, it will cause a loss of power.

Note in Fig. 5 that the oil passages at the rim of the turbine where the oil enters must become smaller as they approach the smaller diameter of the turbine. As the same volume of oil must squeeze through these funnel-like passages, the oil stream will speed up when it leaves the turbine. This speed is used to increase torque by directing it against the stator, which acts as a fluid lever or fulcrum. The stator changes the direction of flow and sends the oil into the pump in the same direction as pump rotation.

Continued on next page

TORQUE CONVERTERS

Let's see how the stator does its job (Fig. 6). A stream of oil aimed at a flat surface splashes off at various angles. The oil can be made to flow more smoothly by curving the entrance and can be reversed by more curving, with a resulting increase in force, indicated by the large arrow.

The stator has curved blades (as in C), which the oil strikes as it leaves the turbine. These blades turn the oil back in the direction of pump rotation.

A—Flat Surface
B—Curved Entrance
C—Curved Blade

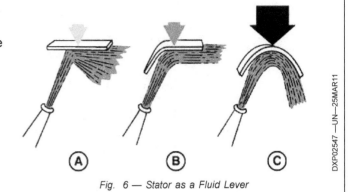

Fig. 6 — Stator as a Fluid Lever

Now that the stream of oil is moving in the same direction but at a greater velocity, it enters the pump smoothly (Fig. 7). Its velocity is added to that developed in the pump so that the total velocity at the pump exits has been increased.

This regenerating action is the key to multiplying torque in a torque converter.

To change the direction of oil flow, the stator must be stationary during the increase in torque.

However, once the pump and turbine are turning at the same speed, it would create resistance. Therefore, the stator is sometimes mounted on a freewheel clutch so that it can turn in one direction only (once torque stops increasing). (In some torque converter designs, the stator may be fixed to the converter case.)

A—Turbine Vane
B—Pump Vane
C—Direction of Engine Rotation
D—Stator Vane

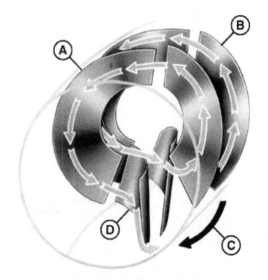

Fig. 7 — Flow Direction Is Reversed by the Stator Vanes

Continued on next page

TORQUE CONVERTERS

DECREASING THE TORQUE

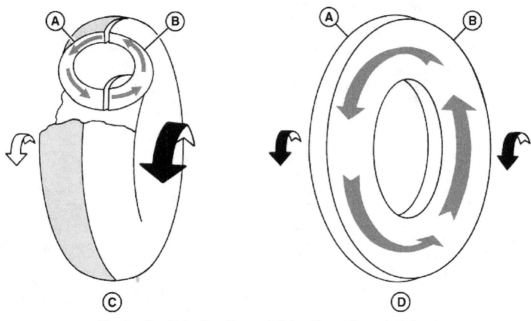

Fig. 8 — How Vortex Flow Changes to Rotary Flow as Torque Is Reduced

A—Turbine
B—Pump
C—While Torque Is Increasing—More Vortex Flow—Pump Is Turning Faster
D—While Torque Is Decreasing—More Rotary Flow—Pump and Turbine Reach Same Speed

Torque is increased as long as the engine is accelerating to get the machine under way. But as the engine speeds up, the turbine also speeds up, which causes the vortex flow of oil in the converter to decrease. At the same time, the rotary flow increases.

Vortex flow keeps on changing to rotary flow (Fig. 8) until the pump and turbine are "locked up" and all torque increase stops.

The torque converter now acts as a simple fluid coupling, sending the same torque it receives on to the drive wheels.

The torque converter is able to automatically reduce or increase torque in infinitely small steps to match the needs of the machine and its driver. This has the same effect as shifting speeds in a gear transmission except that it is done smoothly and automatically while "on the go."

WHAT IS TORQUE CONVERTER STALL SPEED?

Stall speed is the point when a converter has reached its maximum fluid flow and torque multiplication has peaked. Stall speed is determined by the number, the shape, and the angle of the vanes in the impeller, turbine, and stator. By changing the angle of the vanes, stall speed is changed and can allow the engine to run at peak performance. Factors such as torque converter size, clearance between impeller and turbine, and the structural strength of stator vanes determine the amount of stall speed. But at the same time, it can increase the ratio of torque multiplication to transmission. As a rule, the higher the stall speed, the higher the ratio of torque multiplication. Engines vary in horsepower along with the rpm at which they develop their greatest torque and power. Torque converter vane size, angle, and stall speed must all be matched to produce the best performance for the type of application they are to be used for.

TORQUE CONVERTERS

VARIATIONS ON TORQUE CONVERTERS

The table below shows some of the variations in the number of elements used in several torque converters.

Table 1 — Variations on Torque Converters				
Torque Converter Elements	Design A	Design B	Design C	Design D
Pump (Impeller)	2	1	1	1
Stator	2	1	2	1
Turbine	1	2	1	1

The design of a torque converter must match the engine torque and road speed for each application. In off-the-road machinery, torque converters are matched ranging 40 to 600 horsepower (30 to 450 kilowatts). These basic principles we've covered apply to all.

LOCK-UP TORQUE CONVERTER

Automobiles introduced lock-up torque converters years ago to improve fuel efficiency. Even with very light loads, there is always some "slip" in a torque converter. If pump and turbine were turning at the same speed, there would be no vortex flow and no transfer of power.

Slip creates heat and reduces efficiency. To eliminate this loss while keeping all the advantages of a torque converter, it can be equipped with an electric clutch. It locks input and output together for cruising. It unlocks for heavy load or low speed. This feels like one additional shift in the automatic transmission.

TORQUE CONVERTER TRANSMISSIONS

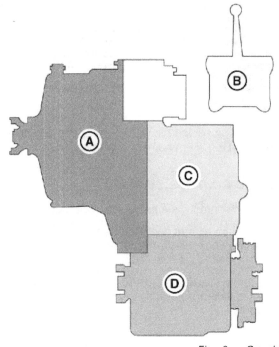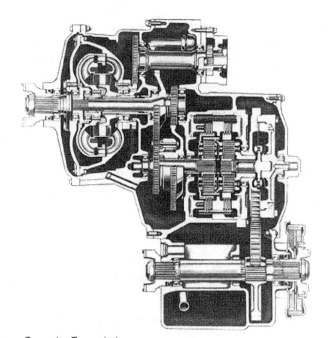

Fig. 9 — Complete Torque Converter Transmission

A—Converter Section
B—Controls
C—Range Gear Section
D—Final Drive Section

The torque converter is but one part of the complete transmission (Fig. 9).

Here are the major components:

- Converter section
- Range gear section
- Final drive section
- Hydraulic control system

Let's look at each part and see what is does for the complete transmission.

Continued on next page

CONVERTER SECTION

We have looked at torque converters with three elements — one pump, one turbine, and one stator.

Now we will examine a twin-turbine model (Fig. 10), which has one pump and one stator, but *two* turbines (first and second).

Each turbine is connected to its individual output gear set. In reality, two converters have been combined into one. The first (blue) turbine is connected to its output shaft by a freewheel clutch.

Here is how the two turbines work together:

When torque demand is high, the freewheel clutch is engaged and the first turbine, assisted by the second turbine, drives the gears. When the machine speeds up, torque demand drops. Then the second turbine takes over the entire load and the freewheel clutch disengages the first turbine.

As a result, the first turbine provides high torque and low speed (for starting up and loading), while the second turbine provides higher speed with lower torque (for travel).

A combining gear set directs torque from the first and second turbines (or the second turbine only) to the range gear section.

RANGE GEAR SECTION

Since torque is reduced and increased automatically in the torque converter, only a few gear sets are normally required in the transmission.

However, as the torque converter rotates in only one direction, it is necessary to have a reverse gear. In some applications, it is also desirable to have low and high range gears as shown in Fig. 10.

Simple planetary gear sets meet the needs of extra gear ranges and lend themselves to hydraulic control.

NOTE: Planetary gear sets and their control by hydraulic clutches are explained in chapter 4.

FINAL DRIVE SECTION

The final drive section includes the transfer drive gear, transfer driven gear, and output shaft (Fig. 10). The output shaft provides for one or two outputs from the same common shaft. Two outputs can be used to propel a four-wheel drive machine as shown in Figs. 9 and 10.

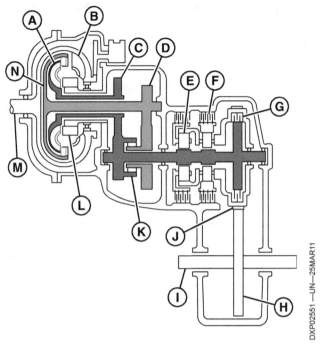

Fig. 10 — Twin-Turbine Torque Converter

A—Second Turbine
B—Converter Pump
C—Second—Turbine Drive
D—First—Turbine Drive
E—Planetary
F—Forward Clutch
G—High Clutch
H—Transfer Driven Gear
I—Output
J—Transfer Drive Gear
K—Freewheel Clutch
L—Converter Stator
M—Input
N—First Turbine

Note also that by adding the transfer gear and one clutch (high range) in Fig. 10, another forward speed range can be obtained.

HYDRAULIC CONTROL SYSTEM

The hydraulic control system uses oil to do these jobs:

- Oil flow lubricates and cools the parts.
- Oil pressure engages the clutches.
- Oil velocity drives the turbines.

Let's use the hydraulic controls for the twin-turbine converter we have just described to see how a typical system works.

Continued on next page

TORQUE CONVERTERS

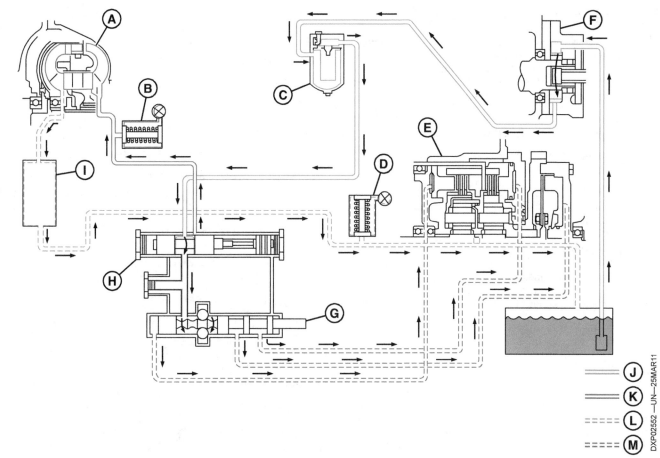

Fig. 11 — Hydraulic Control System for Torque Converter Transmission

A—Torque Converter
B—Converter-Pressure Regulator Valve
C—Oil Filter
D—Lubrication Valve
E—Planetary Clutches
F—Oil Pump
G—Selector Control Valve
H—Main-Pressure Regulator Valve
I—Oil Cooler
J—Oil Pump and Filter Circuit
K—Main-Pressure and Converter-In Circuit
L—Converter-Out Circuit
M—Selector Control Valve Circuit

There are four basic circuits as shown in Fig. 11:

1. Oil Pump and Filter Circuit (shown by blue lines).
2. Main-Pressure Regulator Valve and Converter-In Circuit (shown by red lines).
3. Converter-Out, Cooler, and Lubrication Circuit (shown by dotted blue lines).
4. Selector Control Valve Circuit (shown by dotted red lines).

Let's build up the system and explain each circuit.

OIL PUMP AND FILTER CIRCUIT

Oil is drawn from the transmission reservoir by the oil pump as shown in Fig. 11. The pump delivers its entire output to a full-flow oil filter for cleaning. From the oil filter, the oil supply is sent to the main-pressure circuit.

MAIN-PRESSURE REGULATOR VALVE AND CONVERTER-IN CIRCUIT

The main-pressure regulator valve provides pressure for the planetary clutch packs, directs oil to the selector control valve, and supplies pressurized oil into the torque converter. A converter pressure regulator valve in the converter-in line limits the oil pressure.

CONVERTER-OUT, COOLER, AND LUBRICATION CIRCUIT

The torque converter is continuously filled with oil during operation. Rotation of the converter pump imparts energy to the oil which, in turn, drives the turbines. The oil then flows between the stator vanes, which redirect it to the pump.

Oil flowing out of the converter is directed into the oil cooler. The cooler is a heat exchanger in which the oil flows through water- or air-cooled passages.

From the cooler, oil flows to all passages and outlets in the lubrication circuit.

Continued on next page

TORQUE CONVERTERS

SELECTOR CONTROL VALVE CIRCUIT

Pressurized oil from the main-pressure regulator valve flows into the selector valve bore and surrounds the valve in the area of the detent notches. From this area, main pressure oil is available for operating the low, high, and reverse range planetary clutches.

Moving the selector valve allows oil to change the selected clutch line, to engage the clutch.

This completes the four basic control circuits in our torque converter transmission.

NOTE: For details on hydraulic components, see the "FOS — Hydraulics" manual.

SUMMARY: FEATURES OF TORQUE CONVERTERS

1. Multiply torque.
2. Provide infinite speed ranges.
3. Shift smoothly and automatically.
4. Cushion shock loads on drivelines.
5. Help to dampen vibrations.

TROUBLESHOOTING

INTRODUCTION

As with any transmission, diagnosis is very machine-specific. You must always use the technical manual for the machine you're diagnosing.

However, all torque converters share enough in common to justify a few general comments:

- Any complaints will typically be heat, noise, or power loss.
- Air in the system will create excessive heat and noise. This could be caused by low oil, leaks, or obstructions.
- Contamination can cause catastrophic damage. Do everything possible to keep oil clean.
- If the overrunning clutch for the stator fails to hold, torque will be low.
- If the overrunning clutch for the stator fails to freewheel, top speed will be reduced.
- On a twin-turbine torque converter, the same two comments apply to the overrunning clutch for the first turbine.
- The technical manual will include some pressure checks. This is vital information.

Some diagnostic procedures include a "stall test." Others do not recommend this step. The stall test consists of putting the transmission in gear, locking the brakes, and running the engine at wide open throttle for a few seconds. During the test you check engine speed with an accurate tachometer.

If engine speed is above specification in the stall test, the torque converter probably isn't producing as much torque as it should. If engine speed is below specification, it could be due to engine problems or excessive drag in the torque converter.

The reason others don't recommend using a stall test is that it quickly generates excessive heat. You are putting full engine power into the torque converter and getting zero out. Doing this longer than a few seconds could cause damage.

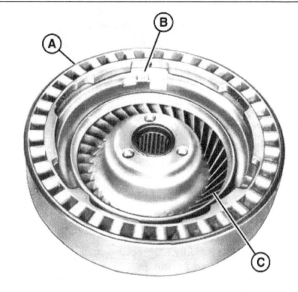

Fig. 12 — Turbines for Twin-Turbine Torque Converter

A—First Turbine
B—Second Turbine
C—Vanes

Oil circulates at high velocity within the torque converter, and any foreign material it carries will rapidly wear down the edges and pit the turbine vanes (Fig. 12), changing their effective shape.

Vane damage will also cause the turbines to become unbalanced. In addition, dirty oil will damage bearings and seals.

Continued on next page

TORQUE CONVERTERS

Some torque converters contain parts made of lightweight aluminum alloys (Fig. 13). Converter housings are usually made of cast aluminum.

Be sure to handle all converter parts carefully to prevent nicking, scratching, and denting.

Parts that fit together closely but with operating clearance will stick if damaged only slightly. Parts that depend upon smooth surfaces for sealing may leak if scratched.

All these parts should be carefully handled and protected during removal, cleaning, inspection, and installation.

Use these rules to help prevent failures:

1. Be sure the oil is kept clean.
2. Service the system at proper intervals.
3. Repair it only if you are qualified.
4. Use all special tools recommended.
5. See the machine technical manual for details.

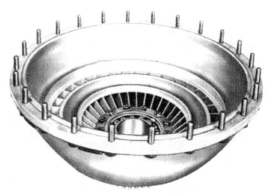

Fig. 13 — Pump for Torque Converter

PRELIMINARY INSPECTION

The operator and service technician can use the eyes, ears, and even the nose to head off trouble before serious damage is done. Here's how:

LOOK:

- Check oil levels. Refer to the operator's manual.
- Check gauges. When starting, running, or stopping, watch the gauges; they tell the story in terms of pressure and temperature.
- Inspect for leaks. Leaks, crimped oil lines, and clogged filters — all affect the converter output.
- Examine the oil. Look for water, dirt, and particles from the converter and the clutch plates.

LISTEN:

- Unusual noises. Listen for squealing from a stuck valve, or grinding or grating sounds from inside the converter.

SMELL:

- Overheating. A strong odor of overheated oil is a major trouble sign. Find the cause at once.

Because one trouble can have the same symptoms as another, only good instruments in the hands of a trained service technician can detect the difference.

Once trouble has been detected in the field, check out the unit with the proper testing tools.

TROUBLESHOOTING PROBLEMS

The troubleshooting given here will cover four common areas of trouble:

1. Overheating
2. Noise
3. Leaks

4. Machine Response

Let's examine each kind of trouble.

1. OVERHEATING

Overheating is a major problem in converter operation. It is affected by the design, the type of work, the operator, the air temperature, and the condition of the unit.

Overheating can cause a loss of power and can damage seals and gaskets and warp metal parts.

A converter may overheat if the work is heavy — not always, but heavier work will generally mean more heat.

If the converter is undersized for the normal work of the machine, it will operate at low efficiency and will tend to overheat. The operator can usually relieve the load on the converter by operating in a lower gear or by reducing the load.

Whenever you receive complaints that a converter in a machine is overheating, try to find out why by watching or by asking whether the correct gears and work methods are being used.

Air in the converter will also cause overheating. Torque converters can work properly only if they are filled with fluid. Air mixed with the oil will cause poor performance, overheating, and possibly serious damage.

Air may enter the system:

1. If the fluid in the reserve tank (if used) is low enough to permit the charging pump to suck air.
2. If a moderately low level of oil permits sucking of air while the machine is working on steep slopes.
3. If there is a leak in the suction line pump gaskets or O-rings (suction leaks may be too small to show up by outside leaking of oil).
4. If the oil or filters are changed or when the lines are opened for any reason.

Continued on next page

6-11

TORQUE CONVERTERS

Summary: Overheating

Let's recap the major causes of overheating in the torque converter.

First, overheating is not only a major problem, it is also a major symptom.

Although normal operating temperatures can be exceeded very rapidly, a machine using a torque converter matched to the job should not exceed its normal temperature when properly used.

Second, the cause of overheating may be found in one or more areas outside of the converter.

Before deciding that the converter is the cause of overheating, check these possibilities:

- Air in fluid system.
- Lack of cooling — plugged cooler core, low coolant level, defective water pump (water-cooled systems only).
- Low fluid level — clogged filter, excessive leakage past converter seals, restricted oil line, defective oil pump.
- Slipping clutches in planetary gear sets.

Third, failure of the converter may cause the overheating. Worn edges and pits on vanes of pump, stator, or turbines will reduce their efficiency.

2. NOISE

Unlike overheating, which can be tested, noise is hard to explain to another person.

A new operator or service technician may never hear a noise directly related to a converter failure.

Yet, an unusual noise heard by an experienced operator or service technician may be the first sign of damage in the converter.

Noise from a converter malfunction may be a whining or growling and may be steady or intermittent.

Worn or dry bearings often produce a hissing noise that will develop into a bumping or thudding sound when they completely fail.

Other sources of noise are worn gears, worn or bent shafts, excessive shaft endplay, shafts misaligned with the engine, and worn freewheel clutches. All these noises mean a possible converter failure.

A mechanic's stethoscope is a valuable aid for isolating noises in the converter.

3. LEAKS

Converter leaks can be one of two types:

- Internal leaks
- External leaks

INTERNAL LEAKS

By internal leaks we mean those within the converter.

As we learned earlier, the converter uses large amounts of oil at high velocity.

If oil is lost from the converter housing through leaks in the pump, turbine, or stator, a loss of power or erratic operation will occur.

Leaks may be caused by the wrong torque on converter bolts.

On some converters, the housing that covers the converter can be removed to determine if a leak has occurred in the converter. Check by starting the engine and operating the transmission until the oil leak shows up.

If leaks form around the converter cover, check the tightness of the cover bolts with a torque wrench. If this fails to correct the leak, disassemble the cover, check the machine surfaces of the cover and flywheel, and install a new gasket.

EXTERNAL LEAKS

By external leaks we mean those occurring outside the converter but still affecting its operation.

These include leaks at cooler lines, filter lines, and pressure or temperature gauge fittings.

Visually check all fittings and oil lines for leakage.

4. MACHINE RESPONSE

Normally a malfunction in the converter will affect the machine's response to load and speed changes.

A machine that lacks power and acceleration at low speed may have a turbine freewheel clutch failure.

Changes in hydraulic pressure, flow, and temperature also affect the performance of the converter, and thus affect the machine's performance.

If you picture the converter as in Fig. 6 (a flow of oil from a nozzle), it is easy to see how heavy cold oil, low pressure, or low flow will affect a response.

TORQUE CONVERTERS

TESTING THE TORQUE CONVERTER

As in troubleshooting, testing is most effective when the engine, converter, and gear train are regarded as a unit, one part affecting the other.

Fig. 14 shows the points where pressures and temperature may be tested on a typical torque converter.

Before testing, several checks should be made.

BEFORE TESTING

1. Check the fluid level. Be sure it is not above or below the required level.

2. Start the engine and warm the engine and transmission to operating temperature.

3. Provide adequate ventilation if starting the machine indoors. Carbon monoxide — a colorless, odorless, and deadly gas — is present in the exhaust of all engines. Be sure you have plenty of good fresh air.

4. Shift into each gear and operate for a minimum of 15 seconds, checking the selector valve detent positions against the related positions on the shift indicator (if equipped).

5. Be sure the PTO is disengaged. Look ahead and to the rear before starting. Make sure no people or obstructions are ahead of you or behind you.

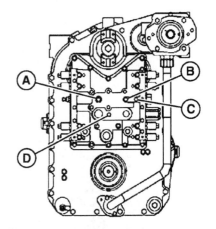

Fig. 14 — Pressure and Temperature Checkpoints on a Typical Torque Converter Transmission

A—Test Converter-Out Oil Temperature
B—Test Converter-Out Oil Pressure
C—Test Main Oil Pressure
D—Test Lubrication Oil Pressure

6. Never allow the transmission to heat up beyond the maximum operating temperature.

TEST YOURSELF

QUESTIONS

1. (Fill in the blanks.) A _____ _____ can only transmit torque, while a _____ _____ can multiply the torque it receives.

2. What part in a torque converter directs the fluid back to the pump?

3. (Fill in the blanks.) Once torque demand is reduced in the converter, the pump and turbine "lock up" and _____ flow is changed to _____ flow.

4. (True or False?) The stator will turn in either direction.

5. (True or False?) Only one turbine can be used in a single torque converter.

6. Match each item on the left with the best phrase on the right.

   a. Oil flow   1. Drives the turbines.
   b. Oil velocity   2. Engages the planetary clutches.
   c. Oil pressure   3. Lubricates and cools.

6-13

INFINITELY VARIABLE TRANSMISSIONS (IVT)

INTRODUCTION

The infinitely variable transmission (IVT) is the newest innovation in power trains. It's the first totally new design in decades. Customer acceptance has been very high, with large numbers choosing the new transmission option.

Automotive manufacturers have also introduced infinitely variable transmissions in recent years.

Benefits include better fuel efficiency, higher productivity, reduced wear, greater flexibility, ease of operation, and a host of convenience features. With manufacturers competing for market advantage, features will vary somewhat from model to model.

IVT can automatically shift up and throttle back under light load. This saves fuel, reduces engine wear, and lowers the noise level.

IVT can automatically shift down under heavy load, adjusting constantly to maintain maximum productivity.

On this model, one lever provides speed adjustment through the entire range from slowest to fastest. The small thumbwheel on the side is used to select maximum speed. Moving the lever in its slot selects a percentage of that maximum.

- A—Set Speed Adjuster (For Setting Maximum Speed)
- B—Speed Control Lever
- C—Set Speed 1
- D—Set Speed 2

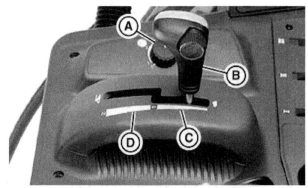

Fig. 1 — Speed Control Lever

Continued on next page

7-1

INFINITELY VARIABLE TRANSMISSIONS (IVT)

On this model, the machine will not move if the lever is held between forward and reverse. No need to use brakes or clutch. This is ideal for loader operation.

- Machine will hold in stopped position.
- No application of foot brakes is necessary.
- Will hold machine and implement stationary on level ground or slopes.

Fig. 2 — Reverser Lever with "Hold" Feature

On this model, the machine will stop if both brake pedals are pushed. No need to use clutch or transmission controls. When the brakes are released, it will resume its previous speed. This is convenient for transport.

- Independent of engine and ground speed.
- Transmission will ratio down to a stop.
- Motion will commence when brakes are released.
- Individual brake pedals for steering brakes, above 1000 engine rpm.

Fig. 3 — Stopping without Clutching or Shifting; Auto Clutch Function

Continued on next page

INFINITELY VARIABLE TRANSMISSIONS (IVT)

On this model, engine speed programming is adjustable by turning a knob. For light loads or transport, Position 3 will throttle back to 1200 rpm under light load while maintaining constant speed. For heavier work, Position 2 maintains at least 1500 rpm. For PTO work, Position 1 maintains standard PTO speed regardless of load.

A—Auto Mode 1 = 2000 RPM (PTO Off); 2100 RPM (PTO On)
B—Auto Mode 2 = 1500 RPM
C—Auto Mode 3 = 1200 RPM
D—Manual Mode

Fig. 4 — IVT Mode Selector

INFINITELY VARIABLE TRANSMISSIONS (IVT)

MECHANICAL OPERATION

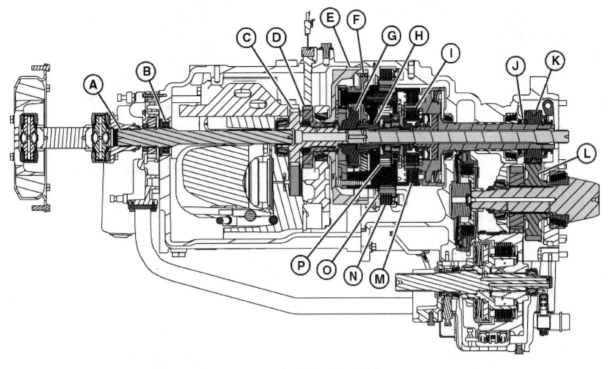

Fig. 5 — IVT with Two Forward Gears

A—Input Shaft
B—Trans Pump
C—Input Gear
D—Hydro Gear
E—Ring Gear
F—Compound Planet Pinions
G—Input Sun Gear
H—High Output Sun Gear
I—High Clutch
J—Clutch Output Assembly
K—Trans Output Gear
L—DDS Gear
M—Low Clutch
N—Reverse Brake
O—Reverse Ring Gear
P—Reverse Sun Gear

Designs vary. Perhaps the simplest version (Fig. 5) has only two forward gears and one reverse, plus the hydrostatic input for variable speed. It has two clutches, one brake, and a compound planetary. This design is the basis for much of the following discussion. Other IVTs have more components, but they are just variations on the same theme.

In basic terms, IVT adds a hydrostatic input and automatic controls to a simple power shift transmission. Shifts are seamless. Speed can transition smoothly from zero to maximum.

Continued on next page

INFINITELY VARIABLE TRANSMISSIONS (IVT)

The heart of the IVT is a compound planetary gear set. It has one ring gear, two sun gears, and compound planet pinions. In Fig. 6, the ring gear has been removed and the front sun gear has been moved to one side to reveal the remaining gears.

The front sun gear turns at engine speed at all times. The ring gear is controlled by a hydrostatic drive. It turns at variable speed in either direction.

The planet pinion carrier turns at variable speed, controlled by the two inputs working together. The planet pinion carrier is the input for low gear and reverse.

The planet pinions are compound gears. The small gear in front and the large gear in back are one piece. The rear sun gear is driven by the large gears on the planet pinions. The rear sun gear is the input for high gear.

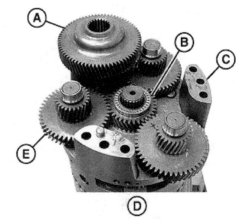

Fig. 6 — Compound Planetary Gears

A—Input Sun Gear
B—High Output Sun Gear
C—Carrier
D—View from Rear of Assembly
E—Compound Planet Pinions
F—Ring Gear (Not Shown)

The sun gear turns at a fixed speed. The ring gear, controlled by a hydrostatic drive, turns either direction at variable speed. The planet pinion carrier is the output.

Fig. 7 shows the simplest situation. The ring gear is not turning at all. The planetary gear set works like a final drive or any simple planetary.

A—Medium Output Speed

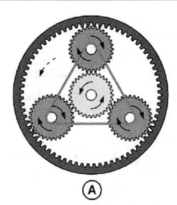

Fig. 7 — Planetary Gear Set with Two Inputs

If the ring gear is turned backward, the output slows down. How much it slows down depends on how fast the ring gear turns backward.

The fastest the ring gear can turn backward is exactly enough to offset the sun gear. The planet pinion carrier doesn't turn at all. This provides the feature shown in Fig. 3, holding the machine stopped without using brakes or clutch. It isn't the same as park, but it will not let the machine roll, even on a slope.

A—Slow Output Speed

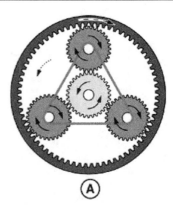

Fig. 8 — Ring Gear Turning Backward

Continued on next page

INFINITELY VARIABLE TRANSMISSIONS (IVT)

If the ring gear is turned forward, the planet pinion carrier turns faster.

The fastest the ring gear can turn forward is exactly the same speed as the sun gear. The entire compound planetary gear set revolves as a unit. The planet pinions do not rotate, because sun gear and ring gear are turning at the same speed. Therefore, the front sun gear and rear sun gear are turning at the same speed. This is the shift point between low clutch and high clutch. It should be a very smooth shift.

A—Fast Output Speed

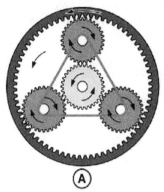

Fig. 9 — Ring Gear Turning Forward

As previously noted, the planet pinion carrier drives low gear and reverse. Fig. 10 shows the carrier removed from the transmission. Compound planet pinions are near the top. Reverse planet pinions are near the center. Notches at the bottom are for a speed sensor. Low clutch and high clutch are inside.

A—Notches B—Carrier Speed Sensor

Fig. 10 — Planet Pinion Carrier

As with any compound planetary gear set, the planet pinions must be properly "timed."

A—Timing Marks B—High Output Sun Gear

Fig. 11 — Compound Planetary Timing Marks

Continued on next page

INFINITELY VARIABLE TRANSMISSIONS (IVT)

Low and high clutches are both in an assembly on the output shaft.

A—Contains High and Low Clutch Assemblies

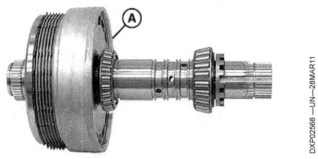

Fig. 12 — Clutches and Output Shaft

The low clutch is the outer portion. When engaged, it locks the planet pinion carrier to the output shaft.

A—Carrier
B—Compensator Piston and Spring
C—Clutch Piston

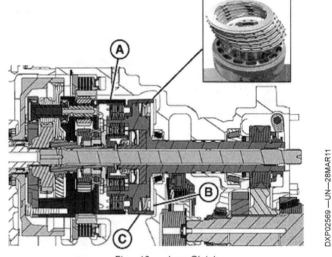

Fig. 13 — Low Clutch

The high clutch is the inner portion. When engaged, it locks the rear sun gear to the output shaft. The sun gear is one end of a hollow shaft. The other end is splined to the high clutch hub.

A—High Output Sun
B—Coil Return Spring
C—Clutch Piston

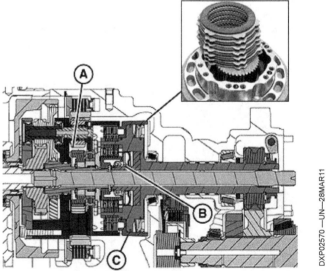

Fig. 14 — High Clutch

Continued on next page

INFINITELY VARIABLE TRANSMISSIONS (IVT)

The reverse brake locks a ring gear to the transmission housing. Because the carrier is turning, the planet pinions walk around the sun gear. The idler pinions turn the sun gear in the opposite direction. (There are three pinion and idler pairs, but only one is shown in the photo.) Though it's a little hard to see in the diagrams, the reverse planetary sun gear is attached to the output shaft.

A—Sun Gear
B—Ring Gear
C—Planet Idler Pinion
D—Planet Pinion
E—Brake Assembly (Disk Shown)

Fig. 15 — Reverse Brake

The output shaft extends from the rear of the planetary gear set. The drilled oil passages are for low clutch engagement, high clutch engagement, and lubrication. A gear on the end transmits power to the differential drive shaft. The gear can be disengaged for transmission testing or for towing a disabled machine. The park brake is on the front of the differential drive shaft. The bottom assembly is the MFWD clutch.

A—Transmission Output Gear
B—DDS Gear
C—Tow Disconnect Collar
D—Transmission Output Gear
E—Output Shaft

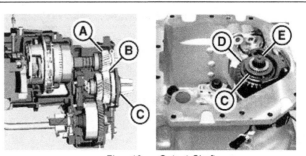

Fig. 16 — Output Shaft

INFINITELY VARIABLE TRANSMISSIONS (IVT)

HYDROSTATIC OPERATION

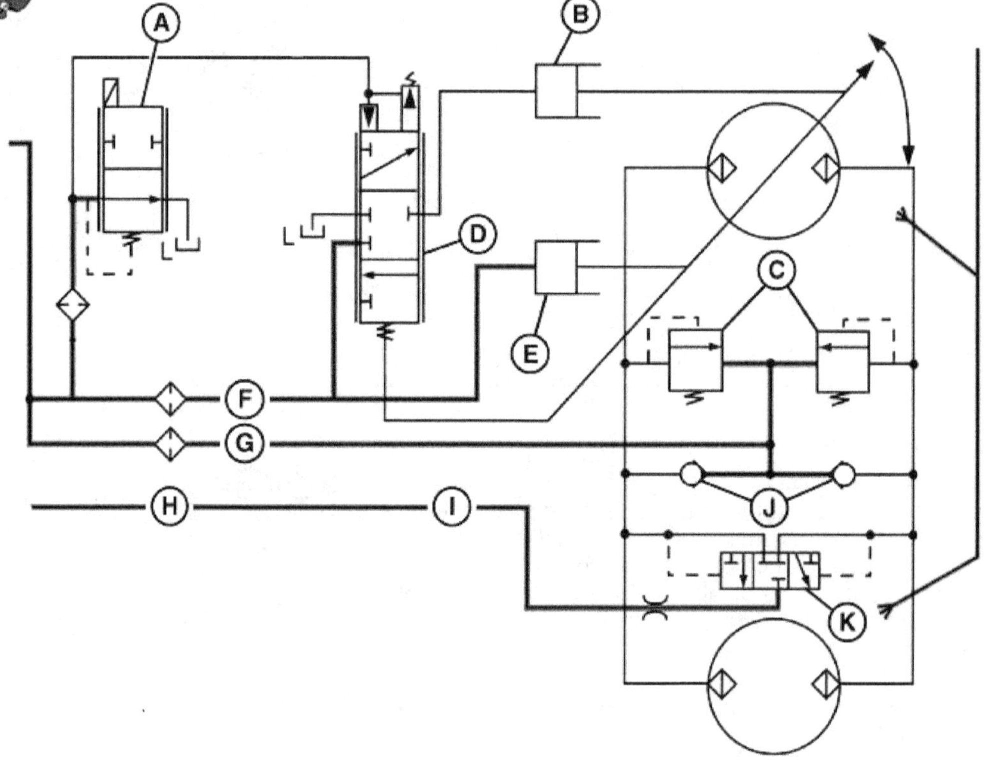

Fig. 17 — Hydrostatic Drive for Ring Gear

A—Hydro Control Pilot Valve
B—Servo Piston
C—High Pressure Relief
D—Hydro Control Valve
E—Bias Piston
F—System 1
G—200µm
H—200µm
I—Loop Flush Return
J—Inlet Check
K—Loop Flushing

Chapter 5 noted the advantages and disadvantages of hydrostatic transmissions. The IVT has all the benefits without the trade-offs. Only one gear in the transmission is driven hydrostatically — not the whole machine. Half of the time, the hydrostatic system is holding back instead of pulling. Efficiency is not an issue.

This particular IVT uses a bent-axis design for hydrostatic pump and motor. Bent-axis components are larger and more expensive than swashplate designs, but they are more efficient over a wider range of speeds. This advantage allows the use of only a two-speed planetary gearbox, because the hydrostatic drive can provide more flexibility.

Continued on next page

INFINITELY VARIABLE TRANSMISSIONS (IVT)

The hydrostatic pump is inside a yoke that swings 45° in each direction from neutral. If the pump is centered, it holds the compound planetary ring gear stationary. By changing the angle of the pump, it can drive the ring gear in either direction at varying speed.

A—Yoke
B—Rotating Group

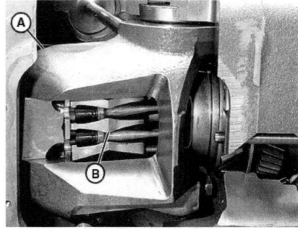

Fig. 18 — Bent-Axis Hydrostatic Pump

Yoke angle is controlled by two pistons, a valve, a cam, and a cam follower. The bias piston always has system pressure. The angle is changed by directing oil to or from the servo piston. This prevents any slack in the system without the disadvantages of using springs.

A—Hydro Control Valve
B—Bias Piston
C—Spring
D—Follower
E—Cam
F—Yoke
G—Servo Piston
H—System 1 Oil
I—From Hydro Control Pilot Valve
J—Sump

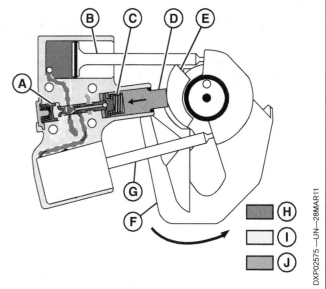

Fig. 19 — Controls for Yoke Angle

Continued on next page

INFINITELY VARIABLE TRANSMISSIONS (IVT)

The hydrostatic motor is held at a constant angle. It turns a gear that turns the compound planetary ring gear.

A—Fixed Unit Gear
B—Rotating Group
C—Ring Gear

Fig. 20 — Bent-Axis Hydrostatic Motor

Operating pressure in a hydrostatic drive can be extremely high. This relief valve is set at almost 8000 psi (550 bar). The same assembly also contains an inlet check valve, which is part of a system to ensure a continuous exchange of oil for cooling and filtering.

A—System 1 Oil
B—Hydro Loop Pressure

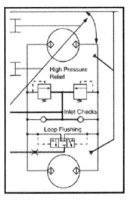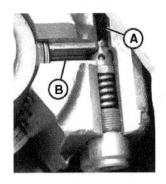

Fig. 21 — Multi-Function Valve

Like fuel injection equipment, this hydrostatic drive system requires specialized facilities and training for service. It must generally be returned to the manufacturer if service is needed.

A—Hydrostatic Drive Module

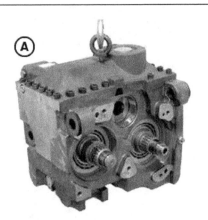

Fig. 22 — Hydrostatic System Removal

Continued on next page

7-11

INFINITELY VARIABLE TRANSMISSIONS (IVT)

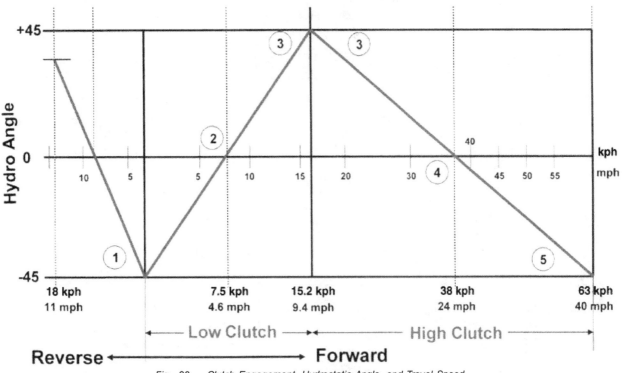

Fig. 23 — Clutch Engagement, Hydrostatic Angle, and Travel Speed

Fig. 23 shows which element is engaged and what hydrostatic pump angle is required throughout the full range of speeds forward and reverse. It is based on 2100 engine rpm. In actual operation, engine speed would typically be lower.

As the machine begins to move forward, low clutch is engaged and the hydrostatic drive is turning the ring gear backward at high speed. As the ring gear slows down, the machine speeds up.

When the ring gear stops, the machine is moving 4.6 mph. (7.5 km/h). Then the ring gear begins to turn forward. Machine speed continues to increase.

When the ring gear reaches maximum forward speed, the machine is moving 9.4 mph (15.2 km/h). Everything in the compound planetary is turning at the same speed — front sun gear, ring gear, planet pinion carrier, and rear sun gear. This is the shift point from low clutch to high clutch.

Now the ring gear begins to slow down. The compound planet pinions turn the rear sun gear, which is now the output, faster and faster.

When the ring gear stops, the machine is moving 24 mph (38 km/h). In the last stage, the ring gear begins turning backward. This transmission is capable of achieving 40 mph (63 km/h), but the control system does not allow such high speed. With reduced engine speed, it still might use that transmission ratio.

INFINITELY VARIABLE TRANSMISSIONS (IVT)

ELECTRICAL CONTROLS

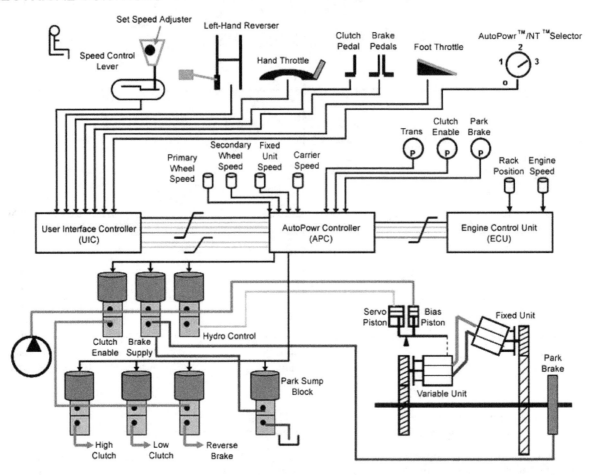

Fig. 24 — IVT Controls

Technology existed 50 years ago for the mechanical and hydrostatic components of infinitely variable transmissions. What has evolved is the control system. Only integrated logic circuits could provide the features of IVT.

Fig. 24 illustrates the eight operator controls, nine sensors, and three computers that control the transmission. Based on the operator's commands and current conditions, the computers send instructions to seven valves. One (clutch enable) modulates engagement pressure. Two are for park brake release. Three are for element engagement (low clutch, high clutch, reverse brake). One is for hydrostatic pump angle.

Continued on next page

INFINITELY VARIABLE TRANSMISSIONS (IVT)

The entire system is "fly by wire." Every control is electronic — hand throttle, foot throttle, speed control lever, etc. Even the clutch pedal is connected to a potentiometer. Like several others, it is a dual-channel Hall-effect sensor. Dual-channel is for redundancy and error detection. Hall-effect is for precision.

Certain motion sensors use dual channels that are 90° out of phase. By noting the sequence in which the two channels turn on and off, the computer can detect the direction of rotation.

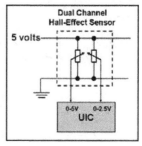

Fig. 25 — Clutch Pedal Potentiometer

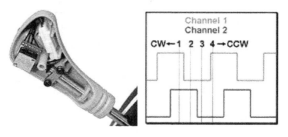

Fig. 26 — Dual-Channel Optical Sensor

Even the engine is included in transmission controls. With light load, the system automatically ratios up and throttles back to obtain the same travel speed at reduced engine speed. This reduces fuel consumption, noise, and engine wear.

Fig. 27 illustrates a typical sequence of events when load increases and then decreases.

1. The operator has set the controls for 20 mph (32 km/h). The throttle is wide open. There isn't much load, so the system automatically reduces engine speed to 1200 rpm while adjusting transmission ratio to maintain the desired speed.

2. As load begins to increase, it takes more power to maintain speed. The system automatically increases engine speed while adjusting transmission ratio for constant speed. Engine speed will increase no more than necessary, but can go all the way to 1950 rpm for maximum power if necessary.

3. If load increases so much that the machine is not able to maintain desired speed, it will do the best it can. Engine speed will hold at 1950 rpm for maximum power. Transmission ratio will self-adjust to prevent lugging the engine any lower than 1950 rpm.

4. As load begins to decrease, the first thing the system will do is return to the desired speed as quickly as

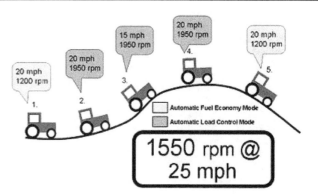

Fig. 27 — Automatic Response to Changing Load

possible. It will maintain 1950 rpm until it reaches the desired travel speed.

5. Then it will throttle back and ratio up to once again maintain travel speed at reduced engine speed.

CREEPER MODE

Incredibly slow creeper speeds are available with IVT. Just by dialing back to minimum speed, you can get down to 164 feet per hour (50 m/h) at wide open throttle.

INFINITELY VARIABLE TRANSMISSIONS (IVT)

COMPARISON WITH ANOTHER POPULAR IVT DESIGN

The IVT design described on the previous pages may be the simplest, but others accomplish the same functions.

Fig. 28 shows a popular version that uses a four-speed planetary gear set. The hydrostatic drive uses swashplates instead of bent-axis design. It has more gears with narrower speed ranges in each, but the end result is the same. The operator would be unable to notice any difference between the two.

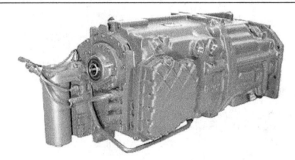

Fig. 28 — IVT with Four-Speed Planetary and Swashplate Hydrostatic Drive

Continued on next page

INFINITELY VARIABLE TRANSMISSIONS (IVT)

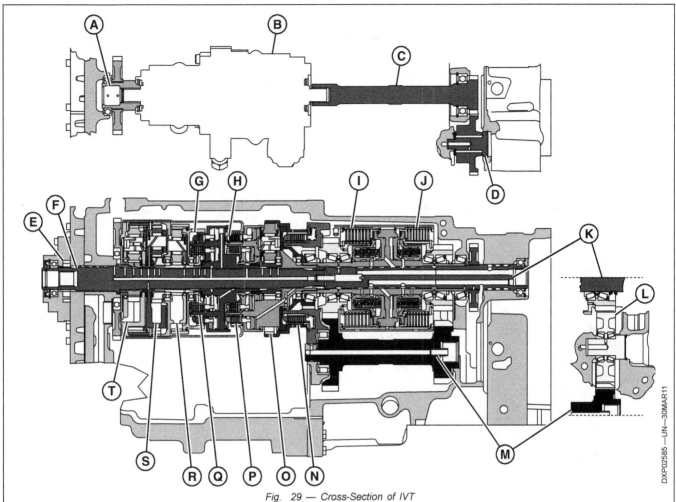

Fig. 29 — Cross-Section of IVT

A—Hydro Output Gear
B—Variable Pump/Fixed Motor
C—Hydro Input Shaft
D—Hydro Drive Gear
E—Transmission Pump
F—Input Shaft
G—C1 Clutch
H—C3 Clutch
I—Forward Clutch
J—Reverse Clutch
K—Auxiliary Shaft
L—DDS Pinion
M—Reverse Countershaft
N—BG/Hi-Lo
O—P4 Planetary
P—C4 Clutch
Q—C2 Clutch
R—P3 Planetary
S—P2 Planetary
T—P1 Planetary

Internal design is completely different:

1. Hydrostatic pump and motor are combined into one assembly, which is mounted alongside the planetaries instead of in front.

2. The engine drives the ring gear, and the hydrostatic motor drives the sun gear instead of vice versa.

3. Forward and reverse are provided by a clutch assembly behind the speed planetaries.

4. There are six clutches and one brake, rather than two and one.

5. Power flows are more complex and difficult to follow.

Continued on next page

INFINITELY VARIABLE TRANSMISSIONS (IVT)

IVT Controls

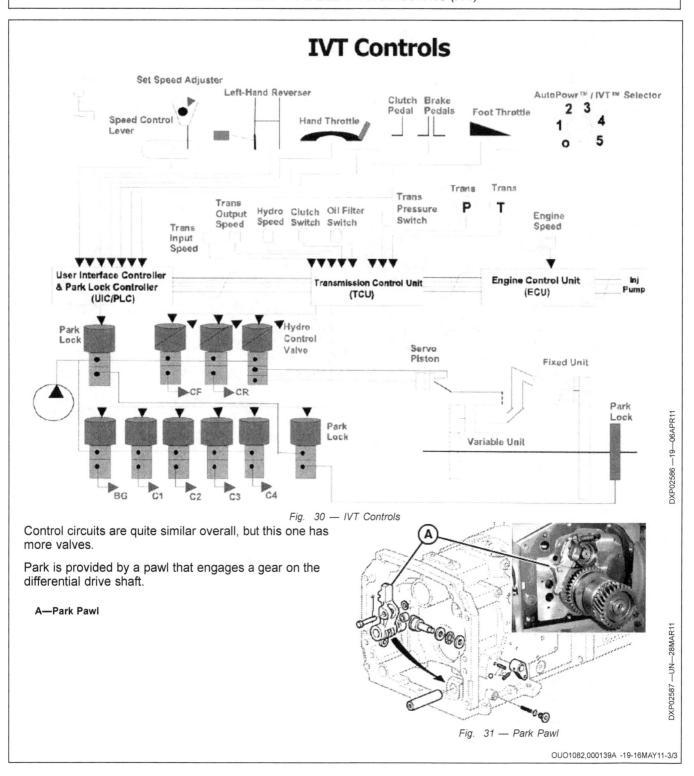

Fig. 30 — IVT Controls

Control circuits are quite similar overall, but this one has more valves.

Park is provided by a pawl that engages a gear on the differential drive shaft.

A—Park Pawl

Fig. 31 — Park Pawl

INFINITELY VARIABLE TRANSMISSIONS (IVT)

SUMMARY

Infinitely variable transmissions are here to stay. They offer unprecedented features that can't be matched by any other design. An IVT is simpler than a power shift, with fewer components. Watch for continuing changes and far more applications.

This overview is only an introduction to IVT. As with any power train, it is very important to use the correct service manual for each specific machine when diagnosing and repairing.

TEST YOURSELF

QUESTIONS

1. The primary reason IVT reduces engine speed under light load is:

    a. To reduce fuel consumption.

    b. To stabilize hydrostatic flow.

    c. To meet drive-by noise regulations in Europe.

2. Suppose a tractor with IVT is pulling a heavy trailer. Transmission controls are set for 20 mph (32 km/h). Throttle is wide open. On a level roadway engine speed drops to 1200 rpm while maintaining 20 mph. If load begins to increase due to an incline, which of the following should occur first?

    a. 20 mph at 1950 rpm

    b. 15 mph at 1200 rpm

    c. 15 mph at 1950 rpm

3. With the same situation as in the previous question, except the hill becomes too steep to maintain the desired speed, which of the following should occur?

    a. 20 mph at 1950 rpm

    b. 15 mph at 1200 rpm

    c. 15 mph at 1950 rpm

4. In terms of mechanical complexity (number of clutches, brakes, gear sets, power flows) is the IVT more or less complex than a power shift transmission?

    a. IVT is more complex than PST.

    b. IVT is less complex than PST.

    c. Both are equal in complexity.

5. Which of the following statements best describes the function of the hydrostatic unit in IVT operation?

    a. The hydrostat is the only connection between the engine and the planetary.

    b. The hydrostat is one of two inputs to the planetary.

    c. The hydrostat is connected to the transmission output shaft, directly controlling direction and speed of tractor movement.

6. Why does the IVT mode selector include one setting in which engine speed remains constant regardless of load?

    a. Steady engine speed provides maximum fuel economy.

    b. Only with constant engine speed is it possible to vary transmission ratio.

    c. Certain jobs, such as PTO work, should be performed at constant engine speed.

7. What makes the shift between low clutch and high clutch so smooth?

    a. Automatic modulation for smooth clutch engagement.

    b. The shift occurs when all parts of the planetary are turning at the same speed.

    c. Engine speed is automatically adjusted to match the oncoming gear.

8. Which operator controls use mechanical linkage to the IVT?

    a. Speed control lever and clutch pedal.

    b. Clutch pedal only.

    c. None — IVT controls are 100% electronic.

9. Certain motion sensors (set speed adjuster, output speed sensor) use dual channels which are 90° out of phase. Why?

    a. To sense direction of rotation.

    b. To improve accuracy by averaging the two readings.

    c. To generate voltage so the system will still function in case battery voltage is lost.

10. Why does one IVT design need only a two-speed planetary gearbox when others need four speeds?

    a. There is no requirement for either approach. It's just designer preference.

    b. Lower engine power requires more transmission speeds.

    c. The two-speed version uses a bent-axis hydrostatic drive.

DIFFERENTIALS

INTRODUCTION

The primary job of a differential is to transmit power 90° to both drive wheels while allowing them to turn at different speeds, if necessary, and still to propel its own load.

Most heavy equipment has a differential lock of some sort, which can be engaged temporarily in poor traction conditions.

RING GEAR AND PINION

Power input to the differential is through the ring gear and pinion (Fig. 1). The pinion is on the end of the drive shaft. The ring gear, sometimes called a "crown wheel," is bolted to the differential assembly.

One function of the ring gear and pinion is to make a 90° bend. The drive shaft runs front-to-rear; the axle runs side-to-side.

Another function is gear reduction. The axle turns more slowly than the drive shaft. The ratio is sometimes optional, so the customer can order whichever is preferred. Other times, ratio is automatically determined by tire size ordered on the machine.

If there were 45 teeth on the ring gear and 13 on the pinion, for example, the reduction ratio would be 3.46:1.

In most heavy equipment, there is one more gear reduction in the final drives, which will be covered in the next chapter.

The ring gear and pinion usually have spiral bevel gears. Gear tooth angle pulls rearward on the pinion under load,

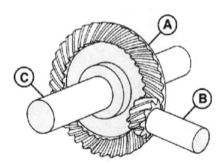

Fig. 1 — Spiral Bevel Ring Gear and Pinion

A—Ring Gear C—Axle
B—Pinion

offsetting forward force from the bevel. If a machine will have frequent heavy loading in reverse, this must be considered when the ring gear and pinion are designed.

Continued on next page

DIFFERENTIAL BEVEL GEARS

The ring gear doesn't drive the left and right axle shafts directly. It turns a "spider" in the center, which has two or three or four bevel pinion gears. The bevel pinion gears are in mesh with two side bevel gears, which are splined to the left and right axle shafts.

When driving straight, both axles turn at the same speed. Bevel pinion gears do not rotate. The complete assembly — ring gear, spider, side bevel gears, and axles — rotates as a single unit. This is illustrated in the top half of Fig. 2.

When the machine is steered left or right, the axles turn at different speeds. This is because the outside wheel travels farther and turns faster. The bottom half of Fig. 2 shows one wheel stopped, as in a brake-assisted turn. The speed difference usually is smaller than this, but it can be any amount.

The ring gear and spider still rotate as before. With the right side bevel gear stopped, the bevel pinion gears rotate to turn the left side bevel gear faster. However much one side slows down, the other side must speed up the same amount. If one wheel is stopped, the other wheel turns at double speed.

It is important to note that power is continuously delivered to both sides. Allowing different speeds will not rob power from either side. Torque is divided equally.

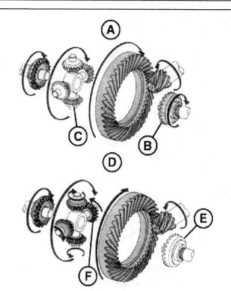

Fig. 2 — Differential in Operation

A—Driving Straight Ahead
B—Side Bevel Gear
C—Bevel Pinion Gear
D—Hard Right Turn
E—Side Bevel Gear
F—Bevel Pinion Gear

DIFFERENTIALS

DIFFERENTIAL LOCK

Allowing the wheels to turn at different speeds can become a problem. Rather than cornering, suppose the difference is due to poor traction. The side with less traction will spin. This reduces productivity, fuel efficiency, and tire life.

Most heavy machinery is equipped with a differential lock; it can be engaged temporarily if one wheel starts to spin. When engaged, it locks left and right sides together for maximum traction. It should be disengaged when not needed.

MECHANICAL LOCK

Smaller equipment usually has a mechanical lock. The operator manually engages a collar that locks two components together.

Fig. 3 shows one type. A hand lever slides a collar on the left axle over to engage splines on the differential carrier. If the left axle is turning at the same speed as the carrier, the right axle is forced to do the same.

It's difficult to see in the illustration, but the two sets of splines extending to the left of the differential are part of the differential carrier. The larger splines are for a parking brake, which isn't shown. The smaller splines are for the differential lock. The axle shaft has a smaller diameter. It extends through the center and into the side bevel pinion.

This type of differential lock may be engaged on-the-go if wheel speeds are the same, but should not be engaged on-the-go if there is much difference in speeds. It's okay to engage before one wheel starts to spin, but not after. The shock loading would be too great.

The differential lock is spring-loaded. It should disengage automatically when both sides have good traction. If it does not, you can tap one brake pedal and then the other to relieve loading so the collar can slide out.

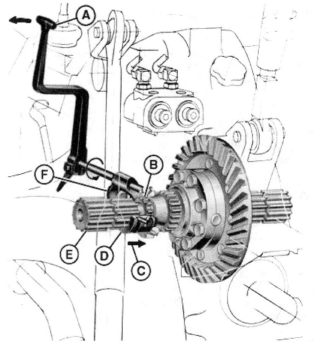

Fig. 3 — Mechanical Differential Lock on a Farm Tractor

A—Differential Lock Hand Lever
B—Carrier Housing Splines
C—Direction to Engage Lock
D—Lock Collar
E—Axle Splines
F—Fork

Continued on next page

DIFFERENTIALS

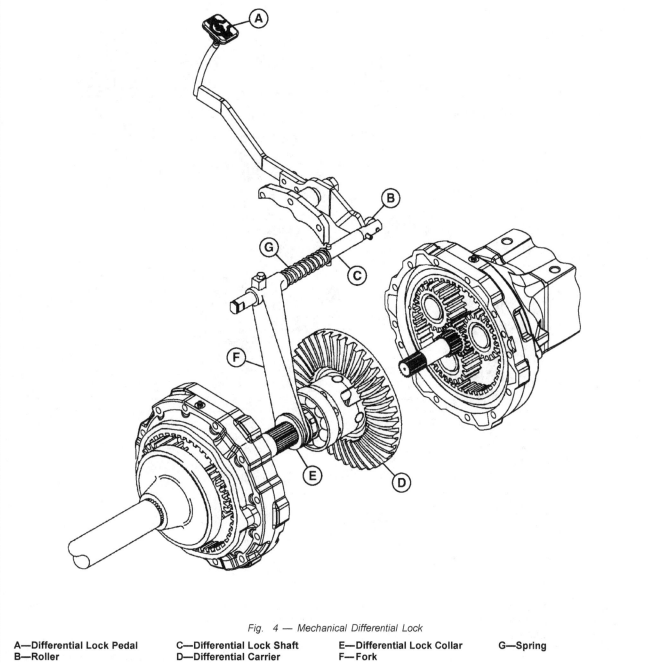

Fig. 4 — Mechanical Differential Lock

A—Differential Lock Pedal
B—Roller
C—Differential Lock Shaft
D—Differential Carrier
E—Differential Lock Collar
F—Fork
G—Spring

Fig. 4 shows a slightly different arrangement. A foot pedal slides the shaft and shifter fork to the right. The collar is splined to the left axle. When slid to the right, pins on the collar engage slots on the carrier.

DIFFERENTIALS

HYDRAULIC LOCK

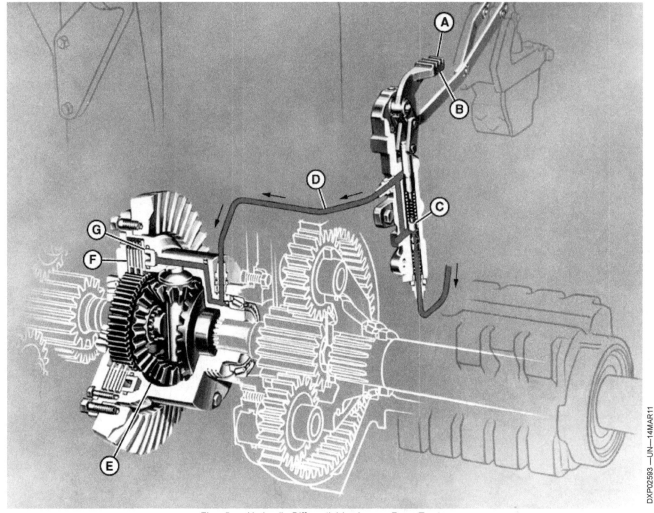

Fig. 5 — Hydraulic Differential Lock on a Farm Tractor

A—Push To Engage Lock
B—Pedal
C—Valve
D—Pressure Oil
E—Bevel Gear (Locked)
F—Disk Clutch
G—Piston Engages Clutch

Larger equipment usually has a hydraulic lock (Fig. 5). When engaged, a valve directs pressure oil to a piston, which squeezes a stack of disks and plates. As before, this locks the left side bevel gear to the differential carrier.

A hydraulic lock can be engaged on-the-go, even if one wheel is spinning.

There is generally some provision to prevent driving with the differential lock engaged all the time. In this example, the control valve is engaged by a foot pedal. If the operator pushes either brake pedal, it automatically disengages the differential lock.

Others use a momentary switch. As soon as the button is released, the differential lock disengages. Some have monitored steering wheel movement, disengaging the differential lock if the wheel is turned more than a certain amount. Computerized controls allow even more options, including multi-function programming.

Continued on next page

LIMITED-SLIP DIFFERENTIAL

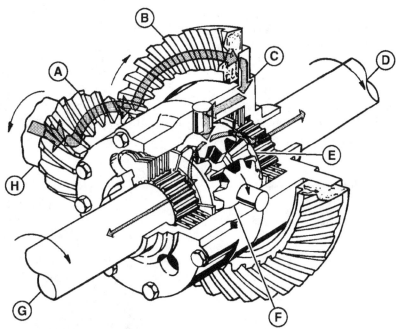

Fig. 6 — Limited-Slip Differential

A—Pinion Gear
B—Ring Gear
C—Differential Housing
D—Axle Shaft
E—Bevel Side Gear
F—Bevel Pinion Gear
G—Axle Shaft
H—Differential Drive Shaft

Fig. 6 shows a design that doesn't really lock the differential, but automatically provides a similar function. It is common in mechanical front-wheel drive axles and automotive applications.

There is a stack of disks and plates on each side. However, the only engagement force comes from the side bevel gear teeth. As the bevel pinion gears drive the side bevel gears, tooth angle also pushes outward. The amount of force against the disks and plates can vary considerably. It is proportional to torque on the axle. Under light load, such as transport, these "clutches" can slip easily to allow differential action. Under heavy draft load, slippage is difficult.

This self-adjustment usually suits the needs, but it has weaknesses in certain situations. If one side was on slick ice, for instance, it wouldn't be able to apply enough torque to prevent spinning.

Continued on next page

DIFFERENTIALS

AUTOMATIC (NO-SPIN) LOCK

Fig. 7 shows quite a different approach. It doesn't have a differential, but does allow one wheel to turn faster when cornering. Neither wheel slows down in a corner. The outer wheel is allowed to freewheel at a faster speed.

The automatic no-spin design is limited to light applications. It does not distribute load to both wheels if they are turning at different speeds. The slower wheel carries the full load, and the faster wheel coasts.

Refer to Figs. 7–10 to see how the no-spin design functions.

A—Driven Clutch and Spider Remain Locked and Travel at Same Speed
B—Spider
C—Driven Clutch Elevated by Cams Disengages from Spider Clutch Teeth and Travels at Faster Speed
D—Splined Side Gear
E—Spider Clutch Teeth Drive Driven Clutch
F—Driven Clutch
G—Splined Side Gear

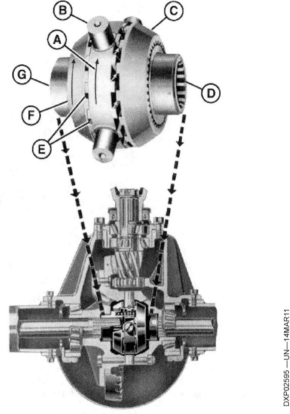

Fig. 7 — Automatic (No-Spin) Differential Lock During a Left Turn

Continued on next page

8-7

DIFFERENTIALS

As with a regular differential, the ring gear is attached to a spider in the center. A driven clutch on each side is splined to a side gear, which is splined to the axle shaft. The driven clutches are spring-loaded toward the center. Square notches on the driven clutches engage mating notches on the spider, driving both axles at the same speed as the ring gear.

Inside the spider is a center cam with tapered notches on both sides. These notches engage mating notches on the cam holdout rings. The cam holdout rings are not splined to the axles, but are driven by a friction fit inside the driven clutches.

Notice in Figs. 8 and 9 how the square notches allow some relative movement between the driven clutches and the spider. If one wheel begins to overspeed, it can turn slightly relative to the spider.

This relative movement enables the tapered teeth on the center cam to push the cam holdout ring outward. The cam holdout ring pushes the driven clutch outward, clearing the square notches. Now the axle and driven clutch can overspeed. The cam holdout ring does not overspeed, but friction against the driven clutch holds it in the extended position.

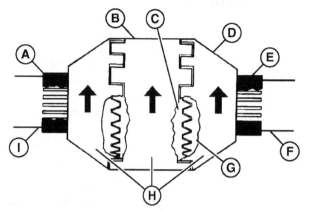

Fig. 8 — Driving Straight Ahead

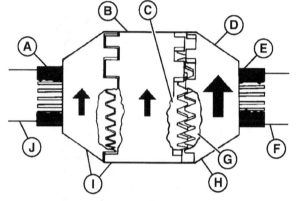

Fig. 9 — Machine Turning Left

A—Side Gear
B—Spider
C—Center Cam
D—Driven Clutch
E—Side Gear
F—Axle
G—Cam Holdout Ring
H—Disengaged – Rotate Faster
I— Locked Together – Rotate at Same Speed
J—Axle

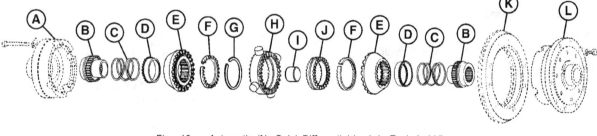

Fig. 10 — Automatic (No-Spin) Differential Lock in Exploded View

A—Differential Housing Cover
B—Side Gear
C—Spring
D—Spring Retainer
E—Driven Clutch
F—Cam Holdout Ring
G—Snap Ring
H—Spider
I— Stop
J—Center Cam
K—Ring Gear
L—Differential Housing

8-8

DIFFERENTIALS

ADJUSTMENTS

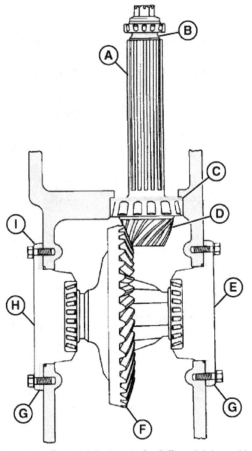

Fig. 11 — Typical Adjustments for Differential Assembly

A—Pinion Shaft
B—Preload the pinion shaft bearings with shims here (step 2).
C—Adjust relation of pinion (depth) to ring gear using shims here (step 1).
D—Pinion
E—Right Quill
F—Ring Gear
G—Preload ring gear bearings with shims under two quills (step 3).
H—Left Quill
I—Adjust backlash of ring gear and pinion by transferring shims between the quills (step 4).

Fig. 11 shows some of the critical adjustments for a typical differential assembly.

1. Fore-and-aft position depth of pinion gear to mesh properly with ring gear.

2. Correct preload on pinion shaft bearings.

3. Correct preload on differential carrier bearings.

4. Correct backlash between teeth on ring gear and pinion.

Always follow precisely the service instructions for the machine being repaired. Such adjustments are very machine-specific, and it's critical to have them right. Incorrect setup can dramatically shorten the life of any power train.

DIFFERENTIALS

DIFFERENTIAL STEERING

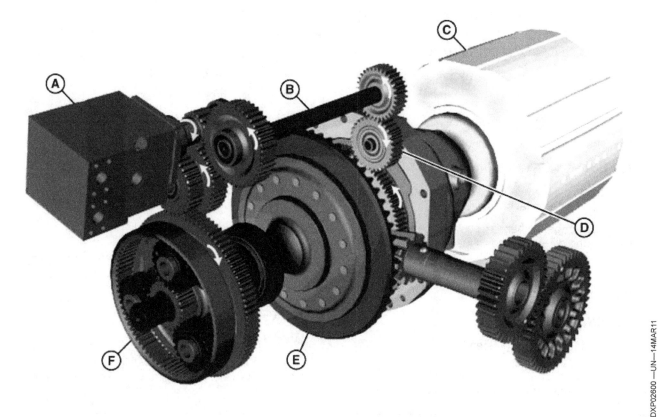

Fig. 12 — Differential Steering for Tracks

A—Steering Motor
B—Steering Cross Shaft
C—Outboard Planetary Final Drive Assembly
D—Reverse Idler Gear
E—Spiral Bevel Ring Gear
F—Inboard Steering Planetary

Tracks machines are steered by controlling the speeds of left and right axles. Some use separate forward and reverse clutches for each axle, controlled by two steering levers.

Fig. 12 shows a differential steering system controlled by a steering wheel. There is no differential in the ring gear carrier. Both sides are locked together. Instead, a hydrostatic motor controls the final drive ring gears for both sides. This will be covered in chapter 9, Final Drives.

DIFFERENTIALS

TROUBLESHOOTING

Here is a guide to the common failures of differentials and the possible causes.

NOISE IN DIFFERENTIAL AT ALL TIMES

1. Incorrect adjustment of ring gear and pinion.
2. Ring gear or pinion damaged.
3. Damaged bearings on pinion shaft.
4. Damaged bearings in differential housing.

DIFFERENTIAL NOT WORKING FREELY ON TURNS

1. Damaged or galled bearing surfaces between bevel gears and differential housing.
2. Damaged or galled bearing surfaces between bevel pinions and differential housing.
3. Damage or galling between bevel pinions and their shafts.

MECHANICAL DIFFERENTIAL LOCK DOESN'T HOLD

1. Linkage between engaging lever and shifting collar broken or improperly adjusted.
2. Shifting collar broken or missing.

HYDRAULIC DIFFERENTIAL LOCK DOESN'T HOLD

1. Oil pressure too low.
2. Blown oil seals or gasket.
3. Failed control valve.

TEST YOURSELF

QUESTIONS

1. What two jobs does a differential do?
2. What are differential locks for?
3. Name the three types of differential locks.
4. Why is the hydraulic lock sometimes linked to the brake pedals?
5. Which type of differential lock is normally engaged?
6. (True or False?) The no-spin differential allows one wheel to turn slower and the other faster than ring gear speed.

FINAL DRIVES

INTRODUCTION

As the name implies, the final drive is the last step in delivery of power from the engine to the wheels.

In almost all farm machines and construction equipment, the final drive includes another gear reduction to reduce speed and increase torque. This gear reduction is not needed in on-highway vehicles.

It would be possible to use a deeper gear reduction in the transmission and eliminate the final drive reduction. However, this would greatly increase torque loading in the transmission, ring gear and pinion, and differential. It is better to run these components at higher speed and lower torque, and then make the final reduction near the wheels.

STRAIGHT AXLES

Cars and light trucks use a direct connection from the differential to the wheels. The ring gear and pinion provide the only gear reduction after the transmission. This is suitable for high speeds and relatively low torque.

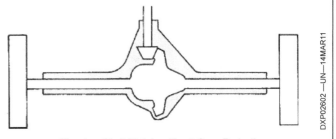

Fig. 1 — Straight Axle without Gear Reduction

Continued on next page

FINAL DRIVES

SEMI-FLOATING AND FULL-FLOATING AXLES

The simplest arrangement has the wheel mounted directly onto the axle shaft (Fig. 2). The axle shaft supports the vehicle weight and all external forces — bumps, braking, cornering, etc. The shaft must be strong enough to withstand all these bending forces in addition to torque. The shaft is supported by a bearing at the outer end of the axle housing. Most cars and light trucks use semi-floating axles.

A—Axle Shaft
B—Bearing

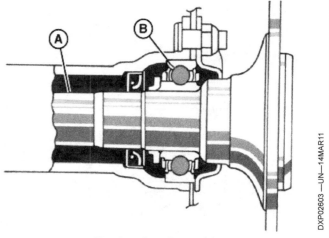

Fig. 2 — Semi-Floating Axle

In a full-floating axle (Fig. 3), the axle shaft merely transmits torque. All external forces are carried by the axle housing. The bearings can be much larger, as they fit over the outside diameter of the axle housing. You could remove the axle shaft with the vehicle resting on its tires. This arrangement is used on most heavy-duty trucks.

One disadvantage of a full-floating axle is that wheel tread is not easily adjustable. This is a big issue on something like a farm tractor, but unimportant in many other types of equipment.

A—Bearing
B—Axle Shaft Only Turns Wheel
C—Drive Wheel
D—Wheel Hub
E—Axle Housing Takes Full Weight of Vehicle
F—Bearing
G—Sleeve Nuts Retain Bearing

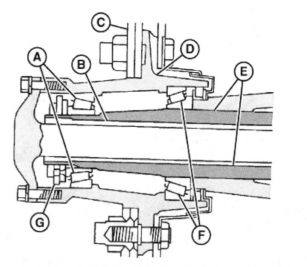

Fig. 3 — Full-Floating Axle, Wheel Hub Supported by Bearings on Axle Housing

FINAL DRIVES

PINION DRIVES

Most farm and construction equipment has some sort of gear reduction in the final drive. One approach is to have a small pinion gear on the shaft from the differential (Fig. 4). The pinion gear drives a much larger spur gear on the axle where the wheel is attached.

Once common, pinion drives are now used mostly for special applications or equipment with lower traction loads. A pinion drive is sometimes referred to as a "bull gear" or "drop axle."

A—From Differential
B—Pinion Gear
C—Axle
D—To Wheel
E—Spur Gear
F—Final Drive Shaft

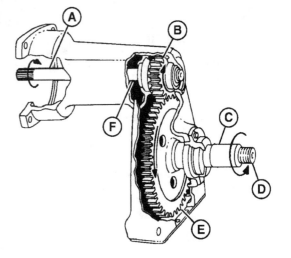

Fig. 4 — Pinion Gear Reduction on Outer Ends of Final Drive

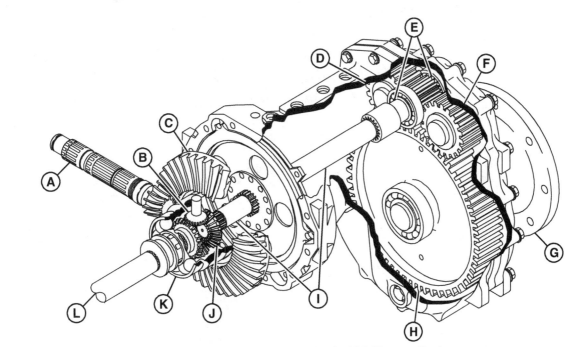

Fig. 5 — Double Pinion Drive for High-Clearance Tractor

A—Differential Drive Shaft
B—Bevel Pinion
C—Differential Ring Gear
D—Idler Gear
E—Pinion Shaft
F—Idler Gear
G—Axle Shaft
H—Final Drive Gear
I—Right-Side Drive Shaft
J—Bevel Gear
K—Differential Carrier
L—Left-Side Drive Shaft

Fig. 5 shows the final drive for a high-clearance machine. The pinion drive provides a large rise from the center of the wheel up to the differential. Notice how it uses two idler gears between the pinion and the spur gear. This cuts gear tooth loading in half, doubling its load capacity.

FINAL DRIVES

CHAIN DRIVES

For an even greater drop, machines such as high-clearance sprayers sometimes use chains and sprockets for the final drive (Fig. 6). Separation of final drive and axle gives axle clearance.

Maintaining proper chain tension is a challenge. An adjustable idler on the slack side of the chain is one option, but that side of the chain is loaded when backing up. It's best to provide adjustable spacing between sprockets.

A—Differential
B—Final Drive Shaft
C—Roller Chain
D—Axle Housing
E—Wheel Hub

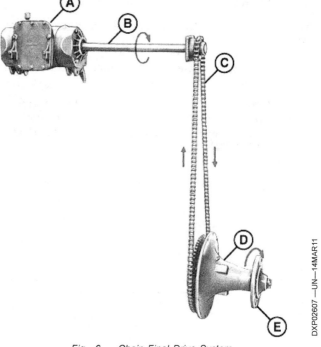

Fig. 6 — Chain Final Drive System

PLANETARY DRIVES

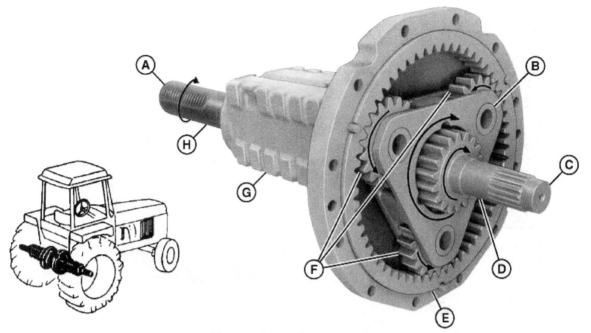

Fig. 7 — Inboard Planetary Final Drive

A—Drive Wheel End
B—Planet Pinion Carrier
C—From Differential
D—Final Drive Shaft and Sun Gear
E—Ring Gear
F—Planet Pinions
G—Rear Axle Housing
H—Rear Axle Shaft

Most heavy equipment use planetary final drives (Fig. 7). The load is spread over multiple gear meshes. Side loading is eliminated by balanced forces on each gear.

The sun gear is driven by the differential. The ring gear is held stationary in the housing. The planet pinion carrier is the output.

Continued on next page

FINAL DRIVES

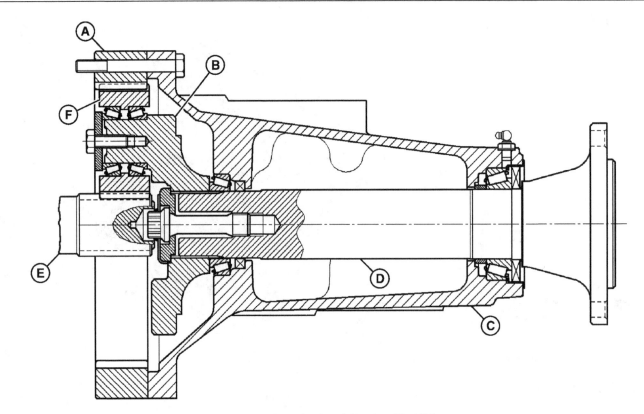

Fig. 8 — Cross-Section of Planetary Final Drive

A—Ring Gear
B—Planet Pinion Carrier
C—Axle Housing
D—Axle
E—Sun Gear (From Differential)
F—Planet Pinion (3 used)

Fig. 8 shows the construction of a planetary final drive. The planet pinion carrier is splined to the axle shaft. The axle shaft is supported by tapered roller bearings at each end. Bearing preload is adjusted by the bolt and retainer that secure the planet pinion carrier.

The outer end of the axle can be flanged (as in Fig. 8) to accept a disk wheel. Or it can extend straight out (as in Fig. 7) to accept an adjustable wheel hub.

Inboard planetaries allow maximum room for adjusting wheel tread.

Continued on next page

FINAL DRIVES

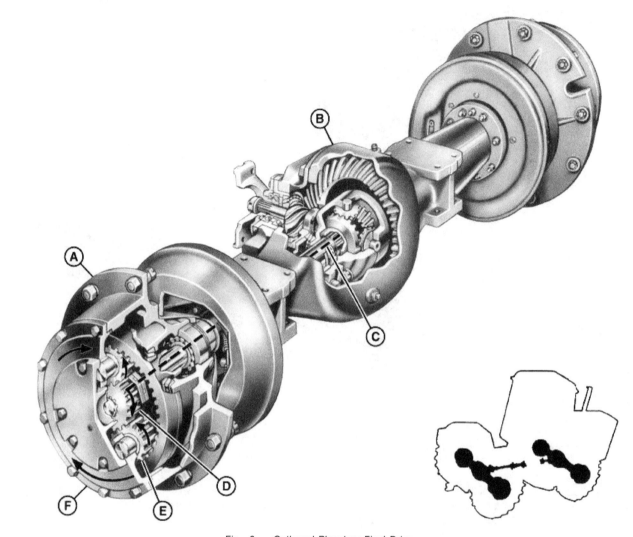

Fig. 9 — Outboard Planetary Final Drive.

A—Ring Gear
B—Differential
C—Final Drive Shaft
D—Sun Gear
E—Planet Pinions
F—Wheel Hub

Outboard planetaries (Fig. 9) can be used if wheel tread adjustment isn't an issue. Operating principles are the same. In this example, the planet pinion carrier is the wheel hub itself.

Also notice that this outboard planetary is a full-floating axle. The hub is carried on tapered roller bearings. The rotating axle shaft does not carry any external loads; it only turns the sun pinion.

The inboard planetaries in Figs. 7 and 8 are semi-floating axles. The axle shaft, bearing, and housing are strong enough to carry the external loads.

Continued on next page

FINAL DRIVES

Fig. 10 — Compound Planetary Final Drive

A—Planet Pinion Carrier
B—Thrust Washer
C—Bearing Rollers
D—Shaft Retaining Pin
E—Set Screw
F—Needle Roller Spacer
G—Planet Pinion
H—Planet Pinion Shaft

Although it's unusual, compound planetaries can be used for greater gear reduction (Fig. 10). Running the transmission and differential at higher speed enables them to carry more power, but that requires additional reduction in the final drives.

FINAL DRIVES

DIFFERENTIAL STEERING

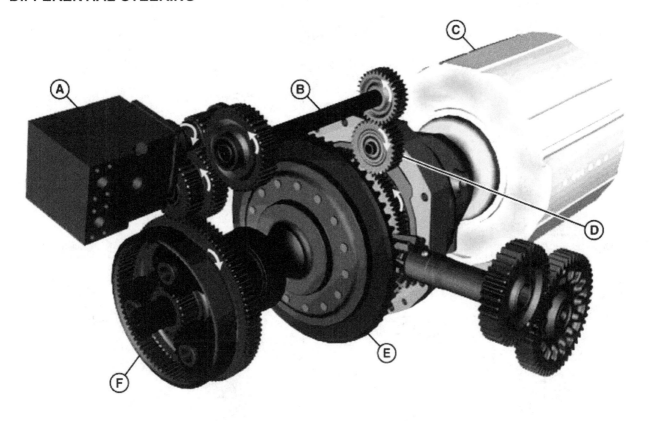

Fig. 11 — Differential Steering in Tracks Final Drives

A—Steering Motor
B—Steering Cross Shaft
C—Outboard Planetary Final Drive Assembly
D—Reverse Idler Gear
E—Spiral Bevel Ring Gear
F—Inboard Steering Planetary

Differential steering is used on certain machines with tracks instead of wheels (Fig. 11). The steering wheel controls a steering motor, which can run in either direction at variable speed. The steering motor turns a steering cross shaft. The planetary final drive ring gears are not stationary as in some other machines. Instead, they can be rotated in either direction by the steering cross shaft. An extra idler gear on one side makes the ring gears turn in opposite directions.

Continued on next page

FINAL DRIVES

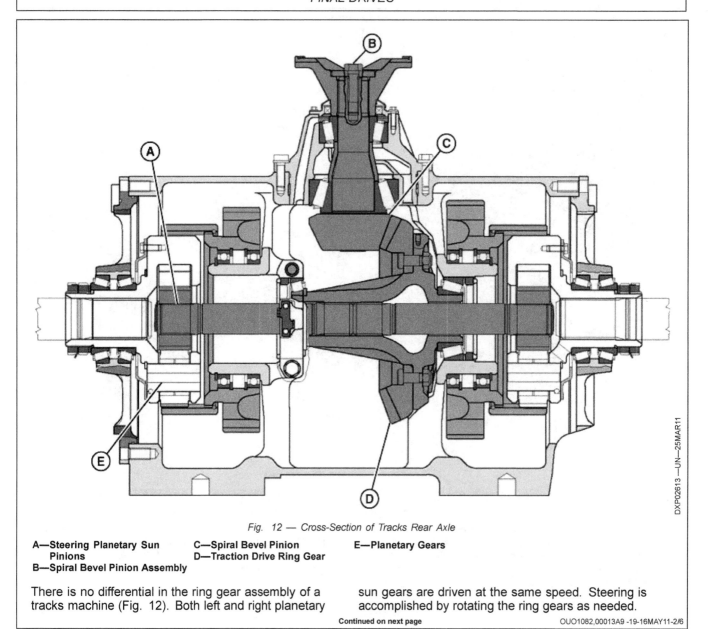

Fig. 12 — Cross-Section of Tracks Rear Axle

A—Steering Planetary Sun Pinions
B—Spiral Bevel Pinion Assembly
C—Spiral Bevel Pinion
D—Traction Drive Ring Gear
E—Planetary Gears

There is no differential in the ring gear assembly of a tracks machine (Fig. 12). Both left and right planetary sun gears are driven at the same speed. Steering is accomplished by rotating the ring gears as needed.

FINAL DRIVES

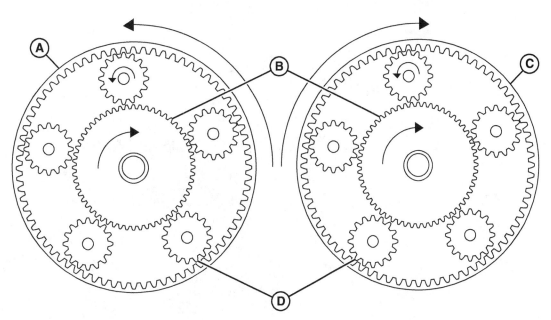

Fig. 13 — Planetary Gears with Two Inputs

A—Steering Ring Gear
B—Steering Planetary Sun Pinions
C—Steering Ring Gear
D—Planetary Gears

As in other planetary final drives, the sun gear is the input. The planet pinion carrier is the output. By rotating the ring gear (a second input, Fig. 13), we control the speed of the output. If the ring gear rotates in the same direction as the sun gear, the planet pinion carrier turns faster. If the ring gear rotates in the opposite direction, the carrier turns more slowly.

When the machine is driven straight, the ring gears do not turn. The planetaries function just like any other final drive.

When the steering wheel is turned, one ring gear rotates forward and the other rotates backward. This speeds one side and slows the other, steering the machine. Turning the steering wheel farther increases the differential, producing a sharper turn.

If the machine is stopped, the steering motor provides the only input. It turns one track forward and the other track backward. The machine pivots in place. Even while driving ahead slowly, it can turn one track backward.

Continued on next page

FINAL DRIVES

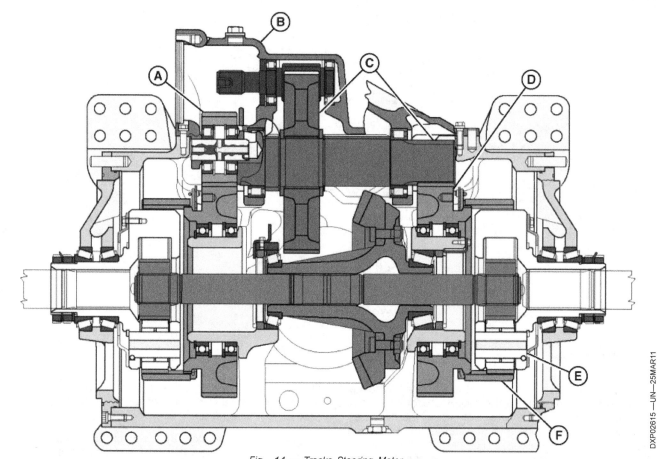

Fig. 14 — Tracks Steering Motor

A—Steering Idle Gear
B—Steering Motor
C—Steering Motor Reduction Gears
D—Steering Ring Gear – External Teeth
E—Planetary Gears
F—Steering Ring Gear – Internal Teeth

Fig. 14 shows a tracks steering motor installation. Unfortunately, it's difficult to trace the power flow because certain parts obscure others. The planetary ring gear on each side is attached to a drive gear.

The steering motor drives both sides. An extra idler gear on one side makes them turn in opposite directions.

Continued on next page

FINAL DRIVES

This design has yet another gear reduction (Fig. 15). An outboard planetary final drive further reduces speed and increases torque. The sun gear is the input (from the differential steering planetary). The ring gear is stationary. The planet pinion carrier turns the hub.

A—Steering Planetary Carrier
B—Planetary Sun Pinion
C—Steering Ring Gear
D—Planetary Gear

Fig. 15 — Outboard Planetary Final Drive

Continued on next page

FINAL DRIVES

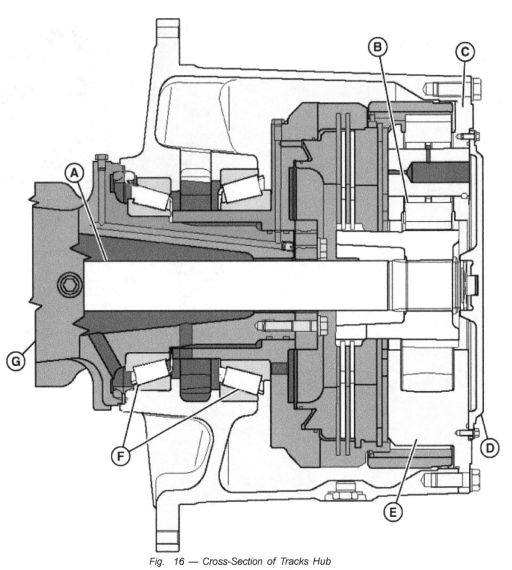

Fig. 16 — Cross-Section of Tracks Hub

A—Outer Axle Shaft
B—Final Drive Sun Gear
C—Outboard Planetary Final Drive Assembly
D—Wheel Hub Cover
E—Final Drive Planetary Carrier
F—Bearings
G—Axle Housing

Fig. 16 looks inside the tracks hub. The outer axle shaft turns the final drive planetary sun gear. The planet pinion carrier turns the outer hub. The sun gear is also splined to the brake disks.

This is another full-floating axle. The hub is supported on tapered roller bearings.

FINAL DRIVES

MECHANICAL FRONT WHEEL DRIVE (MFWD)

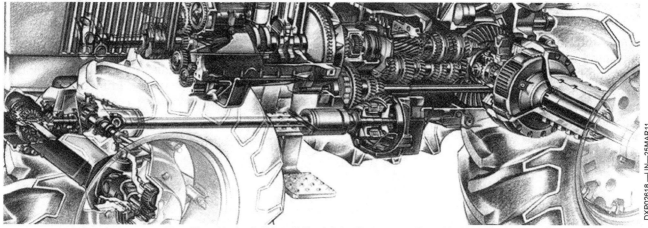

Fig. 17 — Mechanical Front Wheel Drive System on a Farm Machine

More and more machines are equipped with MFWD. It adds traction, reduces load on the rear axle drive, and makes the machine far more versatile.

MFWD is comparable to four-wheel drive (4WD) in a pickup truck. It is usually optional. It is driven by the same shaft as the rear axle, so the speeds are always matched. It can be turned on and off as needed.

The front wheels should have slight overspeed compared to the rears, typically 0%—5%. Whenever the machine is steered around a turn, the front wheels are in a wider circle. Without a bit of overspeed, they would be holding the machine back instead of helping to pull. Even with the overrun, it's helpful to disengage MFWD during sharp turns such as doubling back at the ends of a field.

Continued on next page

FINAL DRIVES

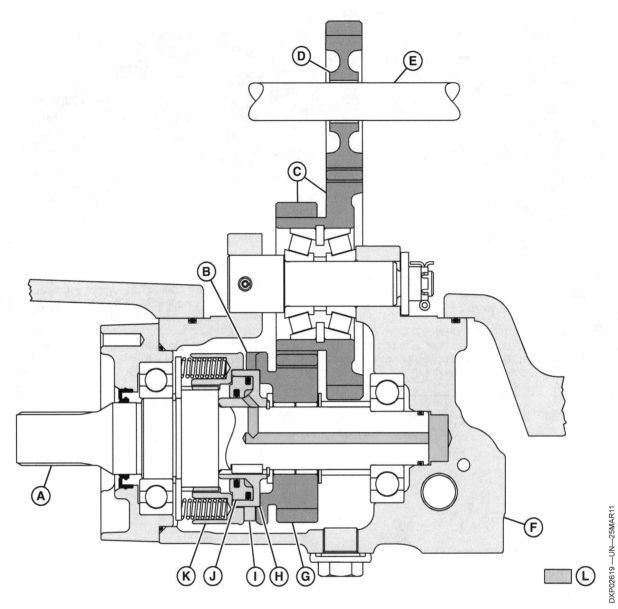

Fig. 18 — MFWD Engagement Collar

A—Output Shaft
B—MFWD Drive Gear Engagement Teeth
C—Idler Gear
D—Drive Gear
E—Differential Drive Shaft
F—MFWD Drop Gearbox
G—MFWD Drive Gear
H—Cylinder
I—Shift Collar Engagement Teeth
J—Piston
K—Shift Collar
L—Return Oil

Various mechanisms are used to engage MFWD. Fig. 18 shows a shift collar controlled by a hydraulic piston. Smaller machines usually have a similar collar controlled by a simple shift lever. Larger machines usually have an electrohydraulically controlled clutch.

Any type can be engaged and disengaged on-the-go if the wheels are not spinning. Those with a shift collar (instead of a clutch) must not be engaged when the rear wheels are spinning, because the shock loading could damage the teeth.

MFWD is always powered by the differential drive shaft, ensuring that front and rear wheel speeds will match. A gear on the DDS drives an idler gear, which drives the MFWD drive gear. If the collar or clutch is engaged, it drives the MFWD drive shaft.

Continued on next page

FINAL DRIVES

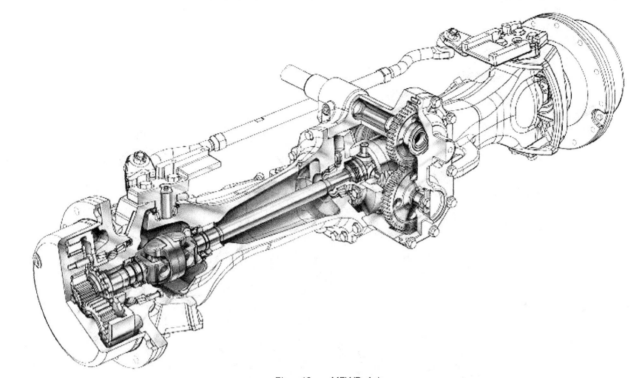

Fig. 19 — MFWD Axle

From there forward, it's a duplication of what we've already seen (Fig. 19). Another ring gear and pinion. Another differential, often with limited slip. Another outboard planetary. Another full-floating hub.

One difference is that the front axle is steerable, so the rotating axle shafts must have universal joints. They are typically constant-velocity joints, to prevent speed pulsations in a sharp turn. The shafts must also have a provision to telescope slightly, as they are shortened in a tight turn.

Continued on next page

FINAL DRIVES

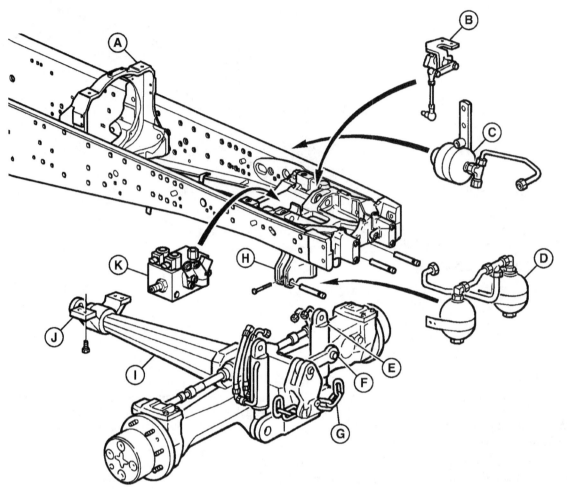

Fig. 20 — Front Axle Suspension

A—Crossbeams
B—Position Sensor
C—Low Pressure Accumulator
D—Accumulator (2 used)
E—Hydraulic Cylinder (2 used)
F—Panhard Link
G—Oscillation Limiter (2 used)
H—Front Axle Carrier
I—Housing
J—Bearing Housing
K—Control Block

Front axle suspension (Fig. 20) is available on some machines. Besides smoothing the ride, this improves traction and reduces stress on the machine. This version uses hydraulic cylinders for support. Accumulators provide an air cushion. A panhard link prevents side-to-side movement. The machine's hydraulic system lets oil in or out to keep the suspension system within the cushioned zone.

Other machines are available with springs instead of hydraulic suspension.

Continued on next page

FINAL DRIVES

Some models are available with independent suspension similar to a car's (Fig. 21). The differential assembly is mounted solidly in the center. The wheels are suspended on upper and lower control arms (B and E). Hydraulic cylinders (F) provide suspension similar to the previous description. Drive shafts (D) have U-joints at both ends.

Manufacturers compete by offering various MFWD features:

- Some engage automatically when both brake pedals are pushed, to provide four-wheel braking.
- Some disengage automatically when only one brake pedal is pushed, to allow tighter turns.
- Some disengage automatically at speeds above a certain point, to reduce tire wear in transport.
- Some engage automatically under heavy load, to reduce stress on the rear axle.
- Some can be programmed into a multi-function switch, so pushing one button will simultaneously do several things.
- Some have a caster angle of 5°–15°, to allow tighter turns by sort of tucking the inside tire under the frame.
- Some have a high pivot point, to prevent tires from striking the vehicle in full oscillation by swinging the high side outward.
- Some have increased steering range, by adding a limited "fifth-wheel" pivot of the whole axle, in addition to the "kingpin" steering on the outer ends.

A—Independent Link Suspension Position Sensor
B—Upper Control Arm
C—Tie Rod Assembly
D—Drive Shaft Assembly
E—Lower Control Arm
F—Suspension Cylinder

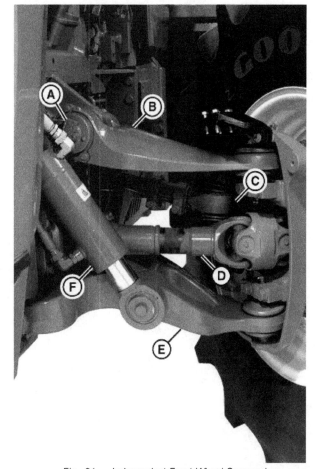

Fig. 21 — Independent Front Wheel Suspension

FINAL DRIVES

ADJUSTMENT

As with any system, it is important to follow repair instructions for your specific machine. Figs. 22–24 illustrate some typical ways to measure bearing end play or preload.

Fig. 22 — Checking Bearing Preload with a Torque Wrench

Fig. 23 — Checking Bearing Preload with a Pull Scale

Fig. 24 — Checking Final Drive Shaft Endplay with a Dial Indicator

FINAL DRIVES

MAINTENANCE OF FINAL DRIVES

The reliability of any final drive depends upon good maintenance, operating at rated load, and proper repair once a failure has occurred.

Watch these key points when diagnosing final drive failures:

1. Excessive drive shaft endplay.
2. Overheating.
3. Lack of lubrication.

EXCESSIVE DRIVE SHAFT ENDPLAY

Excessive drive shaft endplay is normally caused by loose shaft bearings. However, it can be created by:

1. Foreign material in the lubricant, which will wear bearings rapidly.
2. Overloading the machine either by weight or engine torque.
3. Poorly adjusted bearings at the time of assembly.

A continuous noise or knock is a good sign of loose or damaged bearings. On machines with semi-floating axles, such as automobiles, the knocking noise can be heard in the differential case, since the ends of the axle shaft are rapping the spacer block. Readjustment of the bearings that are worn or damaged will not provide a satisfactory repair. Instead, replace the bearings.

Compare the roller ends of a new tapered bearing with a worn bearing. Worn bearing rollers have no shoulder compared to a new bearing — replace them.

OVERHEATING

Many final drives are damaged simply by overheating. This is caused by not maintaining the lubricant at the proper level or by using the wrong type of lubricant.

Galling, pitting, or scoring on the surface of a part indicates the lack of lubricating film and that overheating has occurred.

Overloading and abuse of the machine will also cause overheating. Excessive loads cause deflection in the final drive assembly and concentrate stresses and friction in one area.

LACK OF LUBRICATION

Loss of lubricant through worn or broken oil seals and gaskets may cause severe damage to the final drive. While some bearings are automatically lubricated by oil creeping along the drive shaft from the differential, others are sealed off and require separate lubrication.

When installing a new oil seal, make sure it is flexible or pliable in the area where the seal fits around the shaft.

TEST YOURSELF

QUESTIONS

1. What are the four major types of final drive systems?
2. Which one does *not* give a speed reduction?
3. What is the difference between a "semi-floating" and "full-floating" axle?
4. (True or False?) In a semi-floating axle, the weight of the machine is not carried on the axle shaft.
5. Name one advantage of the planetary final drive over the other three types.
6. What is the main advantage of the chain final drive? The main disadvantage?
7. (True or False?) When the MFWD is engaged, the differential works to automatically balance the driving power.
8. Does differential action occur when power is supplied to the left and right axles?

POWER TAKE-OFFS

INTRODUCTION

Power take-off (PTO) is an auxiliary drive for attached equipment, separate from the traction drive to the wheels. Though it can take other forms, the most common arrangement is a PTO shaft at the rear of a machine. PTOs usually can be engaged or disengaged as needed.

Front-mount or mid-mount PTOs are also available on certain machines. These variations are especially useful for equipment such as snowblowers or belly-mounted mowers.

POWER TAKE-OFFS

TYPES OF PTO

The simplest type of PTO is transmission-driven (Fig. 1, top). An auxiliary gear in the transmission is connected to the PTO shaft. The traction clutch is the only clutch. If you disengage the clutch to stop the tractor, the PTO also stops. This design was common in early tractors. It is still widely used in trucks, where the PTO is engaged intermittently for applications such as a hydraulic hoist pump.

"Live" PTO was a milestone advance in tractor design in the mid-twentieth century. One type is continuous-running PTO (Fig. 1, middle). It incorporates a two-stage clutch — actually two clutches in one assembly, both controlled by one pedal. Push the pedal halfway down to release the traction clutch and stop the tractor. Push it all the way down to stop the PTO. This separation of functions can make a world of difference. You can get the implement up to speed before the tractor moves, and keep it running despite starts and stops. See chapter 2 for information on clutch operation.

Both transmission-driven and continuous-running PTOs also have a shift lever to engage or disengage the PTO by moving a shift collar. The clutch must be disengaged before moving the lever.

Most versatile of all is the independent PTO (Fig. 1, bottom). It has a separate clutch controlled by a separate lever. You can start and stop the PTO while driving. Except for the smallest models, most modern tractors have an independent PTO.

A few tractors have also offered ground-driven PTO (not shown). PTO speed is proportional to travel speed. Engine speed and gear selection are irrelevant. You can even reverse the PTO by backing up. However, driving fast can overspeed the PTO. This design was never widely used.

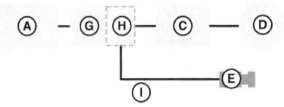
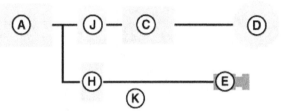

Fig. 1 — Three Types of Power Take-Offs (PTOs)

A—Engine
B—Clutch
C—Transmission
D—Wheels
E—PTO
F—Transmission-Driven PTO
G—Transmission Clutch
H—PTO Clutch
I—Continuous-Running PTO
J—Engine Clutch
K—Independent PTO

POWER TAKE-OFFS

OPERATION

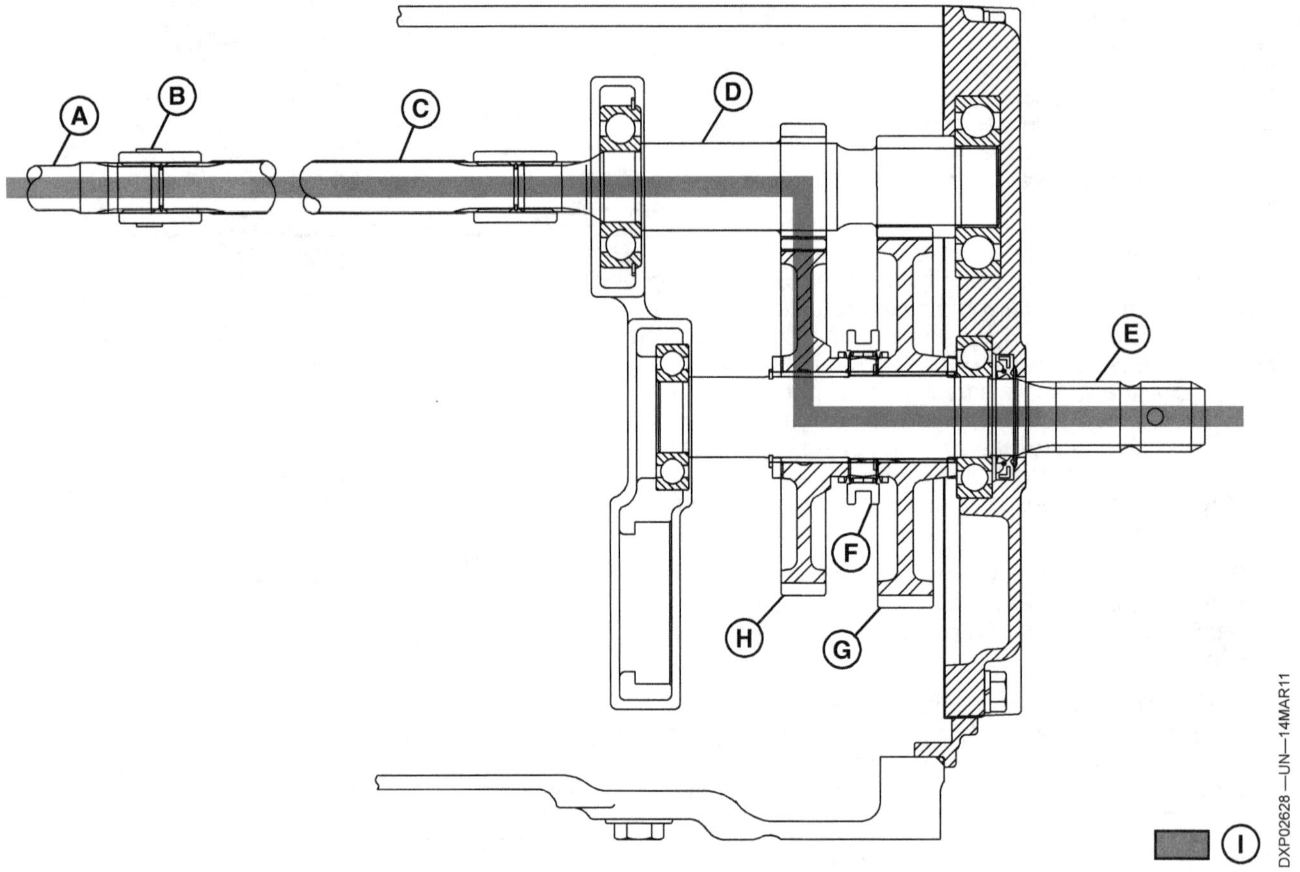

Fig. 2 — Continuous-Running 540/540E PTO

A—PTO Shaft
B—Snubber
C—Intermediate Shaft
D—PTO Pinion Shaft
E—PTO Output Shaft
F—PTO Shift Collar
G—PTO Drive Gear for Standard (540) Mode
H—PTO Drive Gear for Economy (540E) Mode
I—Power Flow

Fig. 2 shows a simple PTO design. It is continuous-running, with a two-stage clutch at the engine flywheel. Unless the clutch pedal is pushed all the way down, the input shaft is turning at engine speed.

This version is shiftable, with two ratios. The PTO pinion shaft has two gears. Each is in mesh with a PTO drive gear on the PTO output shaft. The gears float on the output shaft, so they do not turn the shaft unless the PTO shift collar locks one of them to the shaft. The collar is shown in neutral. It is controlled by a lever on the operator's station.

For heavy PTO loads, the shift collar would be slid back to engage the standard PTO gear. This provides rated PTO speed (540 rpm) at full rated engine speed of 2,400 rpm.

To reduce wear and save fuel on lighter PTO loads, you can shift up and throttle back. Just slide the collar forward to engage the economy PTO gear. This provides rated PTO speed at only 1,700 engine rpm. There is an interlock on the controls to prevent using full throttle and economy mode at the same time. It is important to not overspeed the PTO.

POWER TAKE-OFFS

ASAE STANDARDS

The American Society of Agricultural Engineers (ASAE) long ago established industry standards to ensure equipment compatibility among different brands. Compliance also guards against operating equipment at the wrong speed.

ASAE Standards			
Speed	Diameter	Splines	Power Limit
540 rpm	1-3/8 in. (35 mm)	6	80 hp. (60 KW)
1,000 rpm	1-3/8 in. (35 mm)	21	155 hp. (115 KW)
	1-3/4 in. (45 mm)	20	370 hp. (275 kw)

For years, everything was 540 rpm. As tractor power increased, PTO shaft torque exceeded the limitations of this design. So a second category was introduced. By running the shaft at 1,000 rpm instead of 540, it could transmit 85% more power at the same torque. The 1,000-rpm shaft has different splines to prevent connecting a 540-rpm implement to a 1,000-rpm shaft.

Fig. 3 shows both versions. Manufacturers are free to design the front part any way they wish. The rear part, where an implement is attached, must comply to standards.

Eventually, even more capacity was needed. Now the largest tractors use 1-3/4 in. (45 mm) PTO shaft diameter.

Because customers sometimes connect a small implement to a large tractor, many tractors incorporate

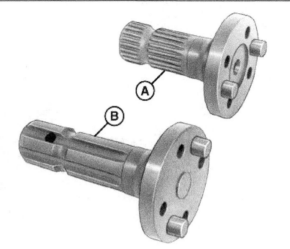

Fig. 3 — PTO Shafts – Two Examples

A—1,000 RPM Stub Shaft (21 Splines)   B—540 RPM Stub Shaft (6 Splines)

two or even three PTO categories. See illustrations on the following pages.

Direction of rotation is always clockwise, as viewed from the rear of the tractor.

ASAE Standards also specify dimensions between the PTO shaft and the hitch point on the drawbar, as shown in Fig. 4. This is to ensure compatibility and prevent damage to the implement's PTO driveline. Most drawbars are adjustable, so it is important to have the drawbar in the correct position when connecting a PTO implement. The drawbar must also be centered behind the PTO shaft and must not be allowed to swing side-to-side.

	PTO-to-Hitch Dimensions
A	14 in. (540 rpm) 16 in. (1,000 rpm)
B	4 in. (101.6 mm) or More
C	6–12-1/2 in. (1-3/8 in. Shafts) [152.4–317.5 mm (35 mm Shafts)] 8–14 in. (1-3/4 in. Shafts) [203.2–355.6 mm (45 mm Shafts)]
D	13–17 in. (1-3/8 in. Shafts) [330.2–431.8 mm (35 mm Shafts)] 17–21 in. (1-3/4 in. Shafts) [431.8–533.4 mm (45 mm Shafts)]

A—PTO Shaft-to-Hitch Point Distance
B—Tractor Tire-to-Hitch Point Distance
C—PTO Shaft-to-Hitch Point Height
D—Ground-to-Hitch Point Height
E—Drawbar
F—PTO Shaft

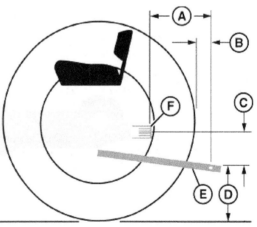

Fig. 4 — Location of PTO Shaft and Drawbar Hitch on a Farm Tractor (ASAE-SAE Standards)

POWER TAKE-OFFS

DESIGN VARIATIONS

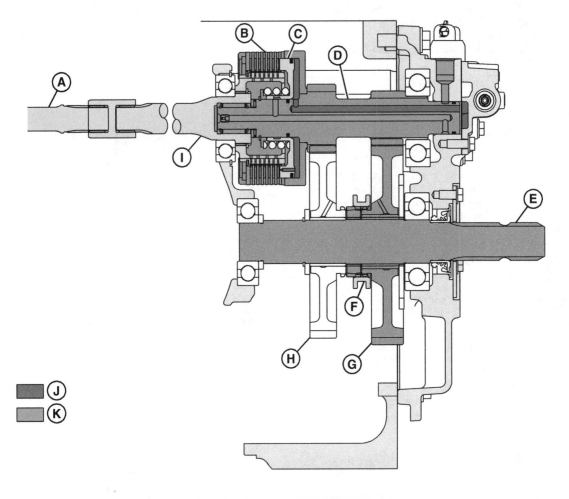

Fig. 5 — Independent 540/540E PTO

A—PTO Shaft
B—PTO Clutch Pack
C—Piston
D—PTO Pinion Shaft
E—PTO Output Shaft
F—PTO Shift Collar
G—PTO Drive Gear for Standard (540) Mode
H—PTO Drive Gear for Economy (540E) Mode
I—Intermediate Shaft
J—Pressurized Oil
K—Lubrication Oil

Fig. 5 illustrates an independent PTO. The PTO clutch can be engaged or disengaged at any time by flipping a switch. An electrohydraulic valve controls the clutch piston.

Like the previous example, this one has 540 standard and 540 economy modes. It is shown with the 540 standard gear engaged by the shift collar.

Continued on next page

POWER TAKE-OFFS

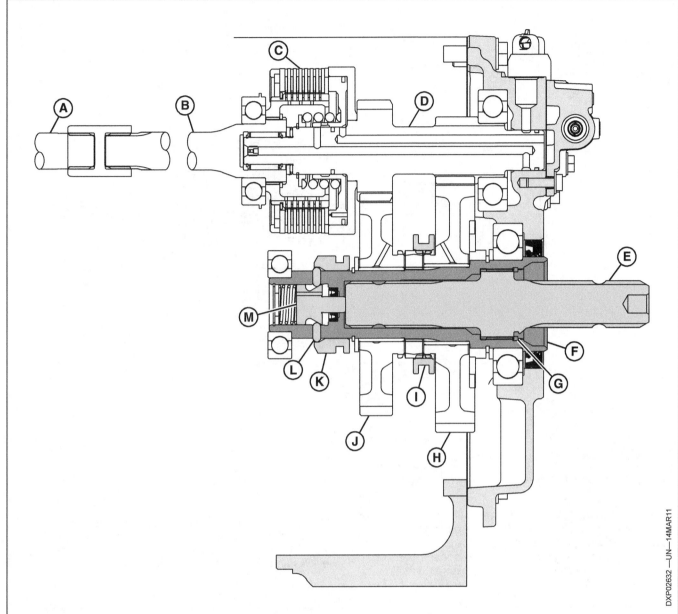

Fig. 6 — Independent 540/1,000 PTO

A—PTO Shaft
B—Intermediate Shaft
C—PTO Clutch Pack
D—PTO Pinion Shaft
E—PTO Output Shaft
F—PTO Output Shaft Receptacle
G—Lock Ring
H—PTO Drive Gear for Standard (540) Mode
I—PTO Shift Collar
J—PTO Drive Gear for Economy (1,000) Mode
K—Lock-Out Shift Collar
L—Lock-Out Pin (2 Each)
M—Lock-Out Cam

Fig. 6 shows an independent PTO with both 540 and 1,000 rpm. To change from one to the other, remove the lock ring, pull the PTO output shaft out of the tractor, and turn it end-for-end.

A lock-out cam prevents installing the shaft in the 540-rpm mode if the PTO shift collar is in the 1,000-rpm position. The lock-out shift collar and PTO shift collar are operated by the same lever. If the PTO shift collar is in the forward position, the lock-out shift collar would hold the lock-out pins in the bottom of the groove on the lock-out cam. This would block the shaft from being fully inserted, and the retaining ring could not be installed. Notice the recess in the opposite end of the shaft. It fits over the end of the lock-out cam, allowing the shaft to be installed in the 1,000-rpm mode.

Continued on next page

POWER TAKE-OFFS

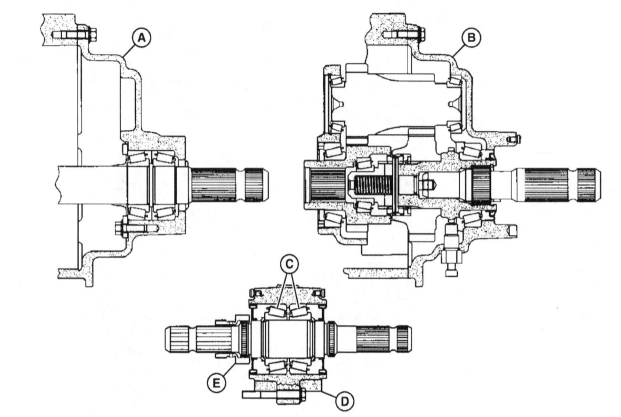

Fig. 7 — 1-3/4 in.(45 mm) PTO and Adapters

A—Standard Rear PTO Output Housing
B—Optional Rear PTO Output Housing
C—Bearings
D—1-3/8 in. (35 mm) Shaft Adapter Housing
E—Torque Limiting Collar

Fig. 7 illustrates 1-3/4 in. (45 mm) PTO on a large tractor. The standard version has only the large shaft. The optional version provides all three categories. To switch, remove the large shaft and install the adapter housing. The adapter housing can be turned in either direction, to provide 540 or 1,000 rpm in the smaller shaft.

Notice the torque limiting collar, which delivers power to the smaller shaft. Similar to a shear pin, it will fail if excessive torque is applied. This is to prevent damaging the PTO shaft or implement driveline.

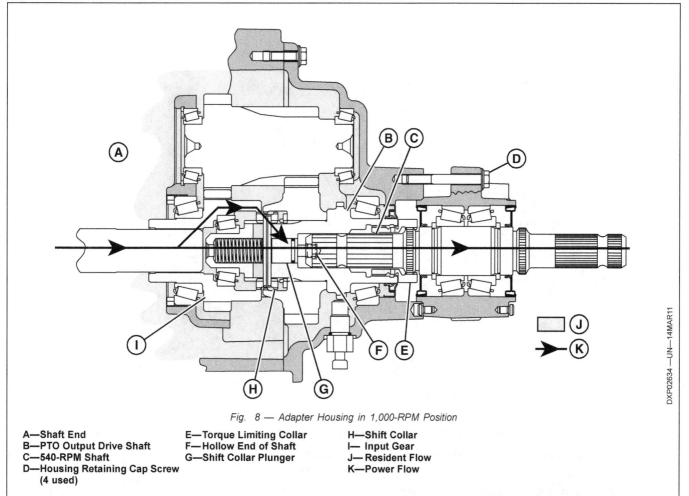

Fig. 8 — Adapter Housing in 1,000-RPM Position

A—Shaft End
B—PTO Output Drive Shaft
C—540-RPM Shaft
D—Housing Retaining Cap Screw (4 used)
E—Torque Limiting Collar
F—Hollow End of Shaft
G—Shift Collar Plunger
H—Shift Collar
I—Input Gear
J—Resident Flow
K—Power Flow

Fig. 8 shows the adapter housing installed in place of the large shaft. It is in the 1,000-rpm position. Notice how the hollow end of the shaft fits over the shift collar plunger. A spring holds the shift collar back. The shift collar locks the input gear to the output drive shaft. The torque limiting collar delivers power to the PTO shaft.

Continued on next page

POWER TAKE-OFFS

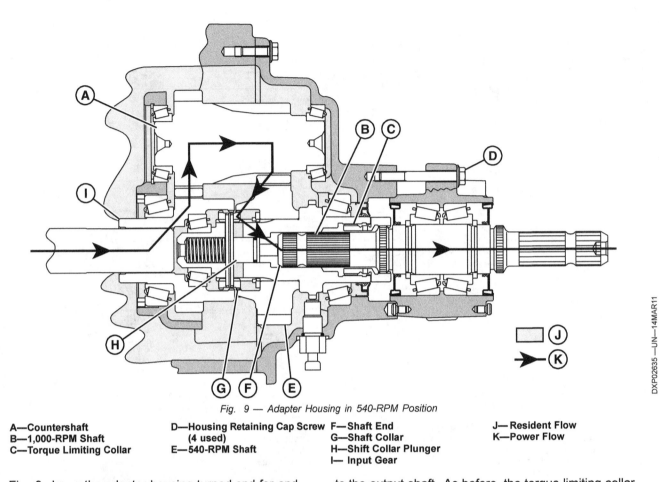

Fig. 9 — Adapter Housing in 540-RPM Position

A—Countershaft
B—1,000-RPM Shaft
C—Torque Limiting Collar
D—Housing Retaining Cap Screw (4 used)
E—540-RPM Shaft
F—Shaft End
G—Shaft Collar
H—Shift Collar Plunger
I—Input Gear
J—Resident Flow
K—Power Flow

Fig. 9 shows the adapter housing turned end-for-end. Now there is no recess in the front end of the PTO shaft. The shaft pushes the shift collar plunger and shift collar forward against spring force. Power flows from input gear to countershaft, to 540 output gear, through the shift collar to the output shaft. As before, the torque limiting collar protects against damage.

It is unlikely that many customers will use 540-rpm PTO on such a large tractor, but the option is there.

Continued on next page

POWER TAKE-OFFS

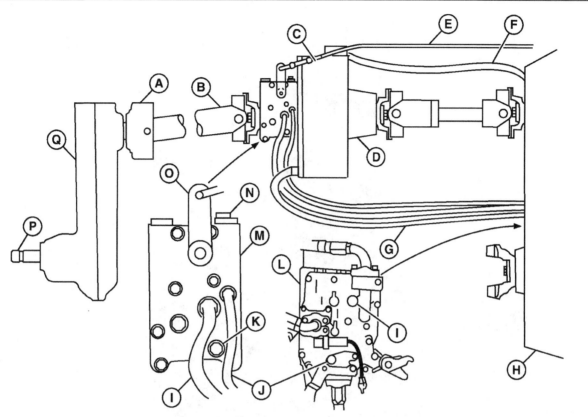

Fig. 10 — Large 4WD Tractor PTO

A—Drive Shaft Flywheel
B—PTO Drive Shaft
C—PTO Clutch
D—PTO Clutch Input Shaft
E—Control Cable
F—Vent Line
G—PTO Gravity Return Hose
H—Transmission
I—PTO Clutch Lube Oil Supply (MST Transmission)
J—PTO Clutch Pressure Supply (MST Transmission)
K—Clutch Pressure Test Port
L—Regulating Valve Housing (MST Transmission)
M—PTO Control Valve Housing
N—Brake Pressure Test Port
O—Control Arm
P—PTO Output Shaft
Q—PTO Drop Gearbox

Many large tractors, especially articulated four-wheel drive models, do not have PTO. Fig. 10 shows a PTO system designed for such tractors. It can be field-installed if needed.

The PTO input shaft (D) is connected at the back of the transmission, to a shaft that turns at engine speed. The PTO clutch (C) controls a shaft (B) connected to the PTO gearbox (Q). Because of the size and power of this tractor, the only PTO shaft (P) available is 1-3/4 in. (45 mm).

POWER TAKE-OFFS

PTO BRAKE

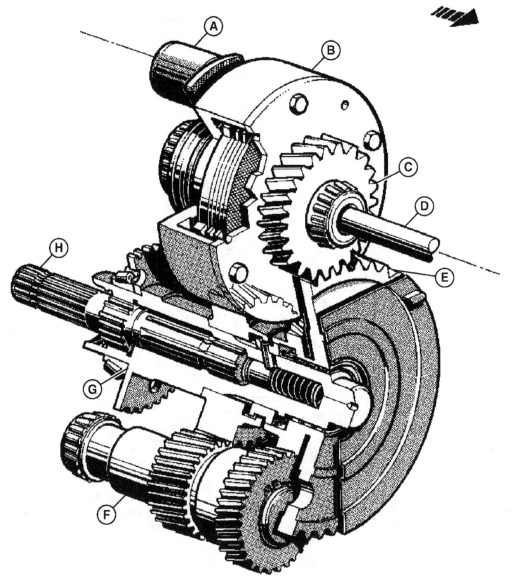

Fig. 11 — PTO Brake

A—PTO Brake
B—PTO Clutch
C—Gear
D—Input Shaft
E—Input Gear
F—Countershaft
G—Output Shaft
H—PTO Stub Shaft

Most PTO systems include some kind of brake to keep the shaft from spinning when the PTO is not engaged. Sometimes it's only a snubber band that drags lightly on the shaft at all times.

Fig. 11 shows a more positive PTO brake. When the PTO clutch is released, pressure oil forces a friction pad against the clutch drum. Whatever the type of PTO brake, it is not designed to stop heavy inertial loads. For implements with high inertia and no overrunning clutch, it's best to throttle the engine down to minimum speed before shutting off the PTO.

POWER TAKE-OFFS

CONTROLS

MODULATION

Engagement of the PTO should be smoothly modulated, especially for heavy or high-inertia loads. Some clutches can be modulated manually by "feathering" the control.

Increasingly, the PTO clutch is engaged electrohydraulically by flipping a switch. The control circuit must include some means to provide smooth engagement. It can be as simple as an orifice to restrict flow of engagement oil and an accumulator to cushion pressure buildup.

Others use computerized controls and pulse width modulated (PWM) pilot valves to operate the engagement valves. These must be calibrated to establish a baseline for the computer.

LOCK-OUT DEVICES

Most tractors incorporate a feature to prevent unintended PTO start-up when the engine is started. On some, it's a neutral start switch to prevent engaging the starter if the PTO is engaged. Others use a spring-loaded control lever that automatically returns to neutral when the engine stops.

Some even include an engagement override device that would block engagement pressure if the engine were jump-started with the control lever engaged. The control has to be in the OFF position with the engine running to shuttle an override valve before the clutch can be engaged. Others accomplish a similar function through electronic circuitry.

FRONT PTO

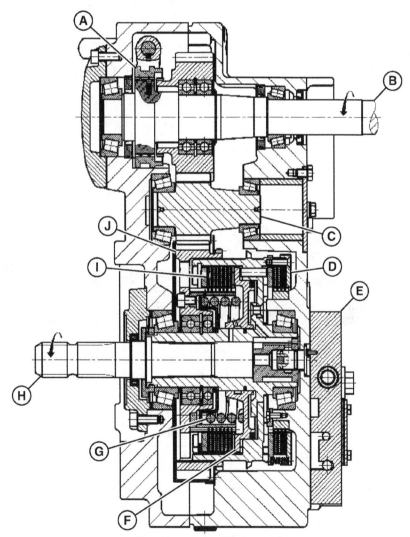

Fig. 12 — Front PTO Assembly

A—Shift Collar
B—Input Shaft
C—Intermediate Gear Shaft
D—PTO Brake
E—Oil Distribution Baffle
F—Ring Gear for Pump Drive
G—Compression Spring
H—PTO Stub Shaft
I—Disk Clutch
J—Ring Gear

Continued on next page

POWER TAKE-OFFS

Many tractor models are available with front PTO. This option, plus a front hitch, can add outstanding versatility. Snowblowers, sprayer pumps, front-mount mowers, spin spreaders, rotary tillers, and all manner of customized equipment can be attached to the front of the tractor.

Especially in Europe, where transport width is very restricted, it has become popular to perform multiple operations in a single pass. Examples include primary tillage in front and secondary tillage in back, final seedbed preparation in front and planting in back, or two mowers at once, with the rear mower swung behind the tractor for transport. Two narrow implements can do as much work as one wide implement.

A front PTO is often installed as a kit, and often produced by an aftermarket supplier instead of the tractor manufacturer. Kits are available for either direction of shaft rotation.

Fig. 12 illustrates one version of a front PTO. The input shaft is connected to the engine crankshaft. This design includes an idler gear to turn the PTO shaft in the same direction as the crankshaft. It has a mechanical shift collar to disengage the drive gear from the input shaft.

The PTO clutch is electrohydraulically engaged. A PTO brake prevents shaft rotation when the clutch is not engaged.

This version has a self-contained hydraulic system. A pump (not visible in the diagram) provides pressure oil for clutch engagement and lube. The control valve is built in. There is even a separate front PTO oil cooler to mount ahead of the radiator.

PTO DRIVES

Proper setup of the implement driveline is very important.

- The vertical and horizontal positions of the drawbar must conform to ASAE specs, as shown in Fig. 4.
- The drawbar must be centered behind the PTO shaft, and must not be allowed to swing side-to-side.
- The driveline must be able to telescope for turns and uneven ground, without risk of bottoming out or disconnecting.
- All shields must be in place and in good condition.
- The driveline must couple securely onto the PTO shaft.
- All universal joints must be in good condition.
- All components must have adequate strength for the application. Different sizes are available to match the load.
- All components must be lubricated according to instructions.
- Never exceed the allowable bend angle.

UNIVERSAL JOINTS

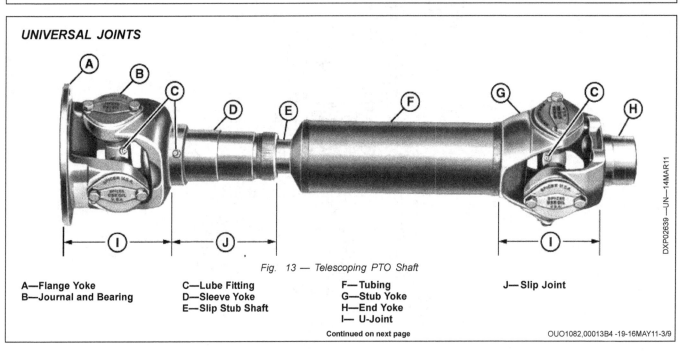

Fig. 13 — Telescoping PTO Shaft

A—Flange Yoke
B—Journal and Bearing
C—Lube Fitting
D—Sleeve Yoke
E—Slip Stub Shaft
F—Tubing
G—Stub Yoke
H—End Yoke
I—U-Joint
J—Slip Joint

Continued on next page

POWER TAKE-OFFS

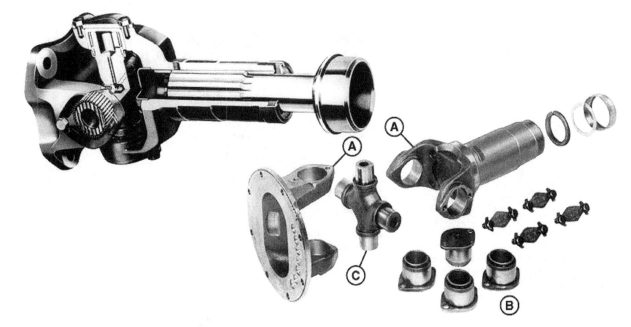

Fig. 14 — Cross and Yoke Style Universal Joint

A—Yokes   B—Bearings   C—Journal

Figs. 13 and 14 illustrate the most common type of universal joint. A bearing cross is installed between two yokes. This allows the shaft to bend in any direction, within allowable limits, while rotating under load. Notice the lubrication fittings for the bearing crosses and the telescoping splines.

PROPER TIMING OF U-JOINTS

Unfortunately, this type of U-joint produces speed pulsations while it is in a bend. With the input yoke rotating at a constant rate, the output yoke speeds up and slows down twice in each revolution. Increasing the bend angle rapidly increases the speed pulsations. This causes severe chatter in the driveline.

If two U-joints are properly "timed," and if both have equal bends in the same direction, the second U-joint cancels any speed pulsations from the first. Only the section of shaft between the two U-joints will pulsate, and this doesn't hurt anything.

Fig. 13 shows two U-joints that are properly timed. The two inner yokes are lined up, both in the same plane. With equal bends in the same direction, this shaft would produce smooth rotation.

If the second U-joint were rotated 90° relative to the first, it would double the speed pulsations instead of canceling them. This could cause failure of the implement and possible damage to the tractor.

Many, but not all, drive shafts have a provision to ensure proper U-joint timing. The telescoping portion may be rectangular, so it can't be inserted 90° out-of-time. If the telescoping portion is splined, it may include a "blind spline," which can be inserted in only one position.

If bend angles are not equal, the second U-joint cannot cancel the speed pulsations from the first. This happens if the front U-joint is near the hitch point and the rear U-joint is much farther back. The front U-joint takes most of the bend, and the driveline chatters in turns. This is also why the drawbar must not swing side-to-side.

On certain implements, smooth operation can be ensured by extending the drawbar hitch point so it is centered between the two U-joints (Fig. 15). This equal-angle hitch is part of the implement. It must be installed on the tractor drawbar when connecting this implement. The length of the drive shaft is adjusted to accommodate this extension.

Fig. 15 — Equal-Angle Hitch

Continued on next page

POWER TAKE-OFFS

CONSTANT-VELOCITY JOINTS

Constant-velocity joints (Fig. 16) are increasingly popular. They produce smooth rotation by incorporating two properly timed U-joints into one assembly. They can accept much sharper bends than one conventional U-joint.

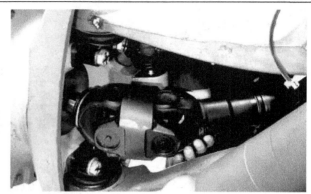

Fig. 16 — Constant-Velocity Joint

A less expensive version is available for low-power applications (Fig. 17). Four large balls, captured in rounded sockets, transmit power.

Fig. 17 — Ball and Socket Style U-Joint

Continued on next page

POWER TAKE-OFFS

SHIELDING

The tractor PTO shaft and implement driveline must be properly shielded. Fig. 18 shows typical shielding on the tractor.

A—PTO Guard   B—PTO Master Shield

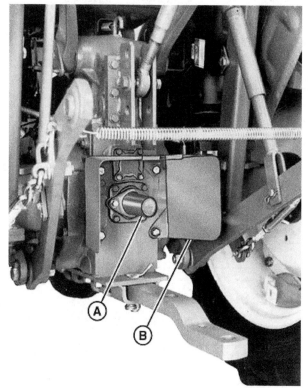

Fig. 18 — Keep Master Shield and Shaft Guard in Place

Early implements usually had stationary "tunnel" shields over the drivelines. Newer implements usually have rotating shields built onto the shafts (Fig. 19). The shield is mounted on bearings. It rotates with the shaft, but can stop if anything contacts it.

Some rotating shields completely surround the U-joints. Far more often, they just have a bell housing as shown in Fig. 19. Together with the master shield on the tractor, this protects against contacting the U-joint.

Every rotating shield should be checked regularly to make sure it hasn't seized to the shaft. With the shaft stopped and the tractor engine shut off, make sure the shield rotates freely on the shaft.

A—Rotating Shield   B—Drive Shaft

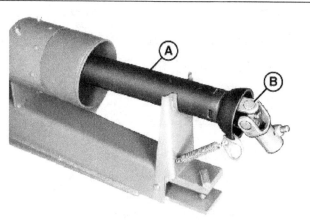

Fig. 19 — Rotating Shield for PTO Driveline

10-16

POWER TAKE-OFFS

SAFETY RULES FOR PTOS

Safety is a prime factor in equipment design.

Here are safety rules for working on machines equipped with PTOs.

1. Keep PTO guards and shields in place, even when the PTO is not operating.
2. Always disconnect the PTO when not in use.
3. Never engage the PTO while the machine engine is shut off.
4. Keep hands, feet, and clothing away from PTO parts.
5. Be sure your clothing is relatively tight and belted. Loose clothing can become caught in moving PTO parts.
6. Never operate PTO shafts at extreme angles.
7. Never ride, or permit others to ride, on the drawbar of the tractor.
8. Never hook 540-rpm equipment to a 1,000-rpm PTO or vice versa.
9. Be sure PTO shields rotate freely at all times. Disengage all power and shut off tractor engine before checking shields.
10. Always be sure PTO drive shaft is properly secured to the machine's PTO shaft.
11. Do not service, lubricate, or perform other work on a machine without first disengaging the machine's PTO and shutting off the engine.

TROUBLESHOOTING PTOS

Troubleshooting PTOs	
Problem	**Possible Cause**
Machine vibrates excessively while opperating	PTO shaft is not properly aligned
Twisted PTO shaft	Overload on PTO shaft rusted from poor lubrication
PTO shafts not telescoping properly	Worn bearings Rotating shields or shaft rusted from poor lubrication
Brinnelling of U-joint journal	Excessive metal fatigue Poor heat treatment
Galling of U-joint journal cross ends	Drive shaft rpm too high
Journal cross end and cups chipping	U-joint capacity too low for required hour and angle
Abrasive corrosion on PTO shaft	Extreme low angle operation

TEST YOURSELF

1. Match each item on the left with the correct definition on the right.

   a. Transmission-driven PTO   1. Operates only when main engine clutch is engaged; stops when disengaged.

   b. Continuous-running PTO   2. Completely separate from main engine clutch.

   c. Independent PTO   3. Can be operated, even though main power train is disengaged.

2. How can you tell two PTO stub shafts — 540 and 1,000 rpm — apart?

3. (True or False?) Never engage the PTO while the machine engine is shut off.

10-17

SPECIAL DRIVES

INTRODUCTION

11

This final chapter briefly covers several power train components not covered in earlier chapters, including:

- Belt drives
- Chain drives
- Reciprocating drives
- Overload protection devices

Refer to Fundamentals of Service—Belts and Chains for more detailed information. This chapter is only a quick overview. Bearings and seals are covered in yet another FOS Manual. See the inside front cover of this book for information on more titles from John Deere Publishing.

SPECIAL DRIVES

BELT DRIVES

Unlike gears and chains, belts are friction devices. A belt transmits power from one shaft to another by way of friction between the belt and the pulleys. A belt can slip. The input/output ratio is not as precise as with gears and chains.

The amount of torque that can be transmitted by a belt depends on several factors:

1. Diameter of the pulleys. Larger diameter gives more leverage, so the same pull produces more torque.
2. Belt tension. A tighter belt gives a tighter grip.
3. Coefficient of friction. The physical properties of both belt and pulley affect the grip between them.
4. Arc of contact. More "wrap" gives more grip.
5. Operating speed, to some degree.

ADVANTAGES OF BELT DRIVES

1. They are simple. They are easy to understand and to service.
2. They are relatively inexpensive.
3. They can absorb shock loading, thus cushioning other components.
4. They can slip to prevent overloading, thus eliminating the need for a slip clutch or shear pin.
5. They can be used as a clutch to engage and disengage drives.
6. With variable pulleys, they can provide speed adjustment.

DISADVANTAGES OF BELT DRIVES

1. They have shorter life, requiring occasional replacement.
2. They have limited load capacity, although large sizes and/or multiple belts can overcome this. In a large combine harvester, for instance, both the traction drive and the machine drive may be belts.
3. They slip or creep slightly in use, meaning they generally can't be used where shafts must stay precisely "timed" relative to each other. However, cogged belts can overcome this, even for something as critical as camshaft timing in an engine.
4. Although most belts just connect two pulleys on parallel shafts turning the same direction, they can offer outstanding flexibility. Fig. 1 shows some of the possibilities.

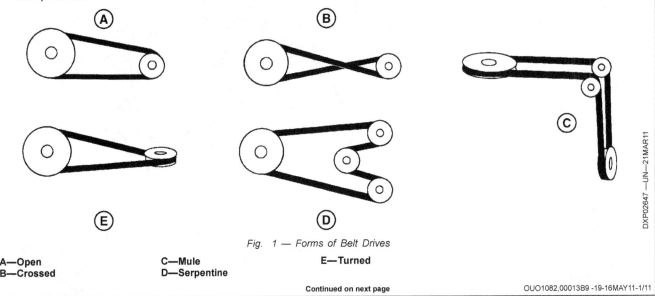

Fig. 1 — Forms of Belt Drives

A—Open
B—Crossed
C—Mule
D—Serpentine
E—Turned

Continued on next page

SPECIAL DRIVES

FLAT BELTS

Many years ago, flat belts (Fig. 2) were widely used to drive stationary equipment such as threshing machines, balers, and sawmills. These belts were long and wide. Power came from a stationary engine or a parked machine, which might be 20 paces away.

Pulleys for flat belts were slightly barrel-shaped, helping the belt center itself. Even so, it was a challenge to keep the belt in position and to stay out of its way.

Such belts have all but disappeared, replaced by much more convenient and much safer equipment.

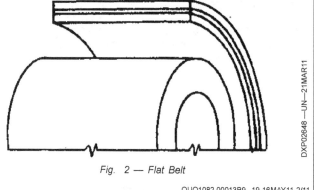

Fig. 2 — Flat Belt

V-BELTS

V-belts (Fig. 3), on the other hand, are used in countless applications. Deep grooves in the pulleys hold the belt securely in position. A V-belt can pull much harder with less tension than a flat belt, because the belt is squeezed between the sides of the pulley.

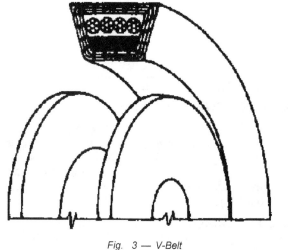

Fig. 3 — V-Belt

For greater capacity, multiple belts can be installed in parallel (Fig. 4). Multi-rib belts are increasingly popular, combining the convenience of a single belt and the capacity of several.

For applications with working pulleys (not just idlers) on both sides of the belt, hexagonal belts provide a "V" on each side.

Fig. 4 — Drive Belt Using Multiple Belts

Continued on next page

SPECIAL DRIVES

ROUND BELTS

Round belts (Fig. 5) exist, but they are rare on anything much larger than a sewing machine.

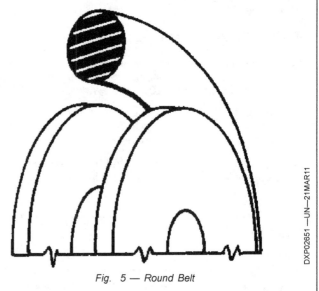

Fig. 5 — Round Belt

VARIABLE-SPEED BELT DRIVES

A belt drive can function as a transmission by varying the width of one or more pulleys (Fig. 6). Spreading the sides of a pulley (left) allows the belt to move closer to the center, effectively reducing the diameter. Squeezing the sides closer together (right) forces the belt out to a larger diameter.

A second variable is required to accommodate changes in length. This can be a movable idler or a second, spring-loaded, variable pulley.

Either alone or in combination with a range transmission, variable-speed belt drives are used on a wide array of self-propelled equipment.

A—"Smaller" Pulley Turns Faster
B—"Larger" Pulley Turns Slower
C—Faster Speed – Pulley Halves Are Apart and Belt Rides Deep
D—Slower Speed – Pulley Halves Move Together and Belt Rides High

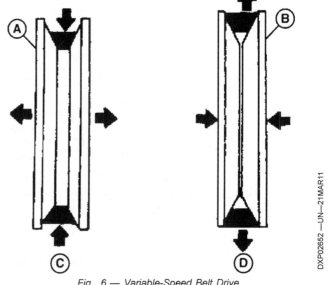

Fig. 6 — Variable-Speed Belt Drive

Continued on next page

SPECIAL DRIVES

BELT TENSION

Too little tension causes slippage, heat, and premature failure of belts. Too much tension overloads belts, bearings, and possibly shafts.

Although belt tension gauges are available, tension is usually specified in terms of deflection (Fig. 7). Measuring in the center of an open span, applying a certain force should deflect the belt a certain distance.

Temperature changes and wear can affect belt tension. Maintaining proper tension is important.

Automatic belt tensioners are increasingly popular. Besides the convenience, they also improve performance and help prevent failures.

Fig. 7 — Checking Belt Tension

BELT DRIVE MAINTENANCE

The outer edge of a V-belt should be approximately flush with the outer edge of the pulley (Fig. 8). If the belt slips deep into the pulley, one or the other is probably so badly worn it needs replacing.

A V-belt should never contact the bottom of the pulley groove. If it does, or if the belt extends well beyond the outer edge of the pulley, belt width is probably mismatched.

A—At Rest
B—±1/16 in.

C—At Rest

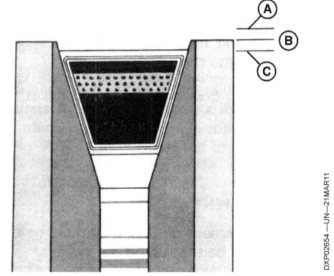

Fig. 8 — Proper Tension of V-Belt in Standard Sheave

Continued on next page

SPECIAL DRIVES

Fig. 9 shows a badly worn pulley, probably the result of abrasive dust. In some environments, pulleys need to be replaced routinely. Installing a new belt onto badly worn pulleys will lead to poor performance and rapid belt wear.

A—Dished Out

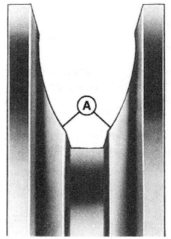

Fig. 9 — Sheave Grooves "Dished Out" by Wear

Pulley alignment is important to belt life. Fig. 10 illustrates severe misalignment. Fig. 11 shows how to check it.

A—Tie Cord to Shaft **B—Cord Must Touch Sheaves at Arrows**

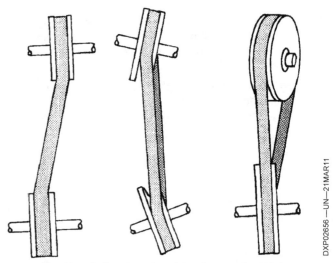

Fig. 10 — Pulleys Not Aligned Can Damage the V-Belt

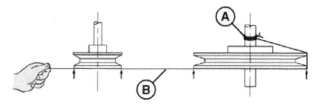

Fig. 11 — Checking Alignment of Belt Pulleys

Continued on next page

SPECIAL DRIVES

Fig. 12 shows damaged pulleys. Fig. 13 shows a wobbling pulley, which could result from damaged pulley, shaft, or bearings.

Exposure to oil, grease, or fuel can cause premature failure of belts. So can excessive heat or prolonged exposure to sunlight.

For more extensive information, refer to Fundamentals of Service—Belts and Chains.

A—Chipped   B—Bent

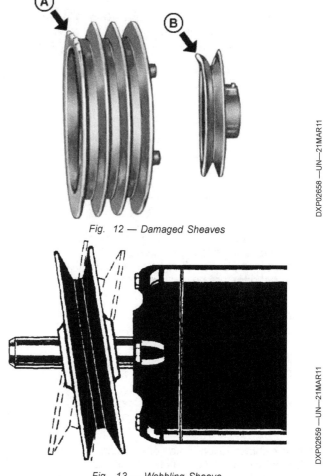

Fig. 12 — Damaged Sheaves

Fig. 13 — Wobbling Sheave

CHAIN DRIVES

Chain drives (Fig. 14) offer most of the advantages of belts. Chains do not creep or slip, which is important in certain applications. Chains can carry greater loads. Chains are not damaged by temperature, sunlight, or petroleum products.

On the minus side, chains require more precise alignment. Chains require lubrication, and they tend to be noisier.

The idler should be on the slack side of the chain. Whenever it's practical, the slack side should be on the bottom.

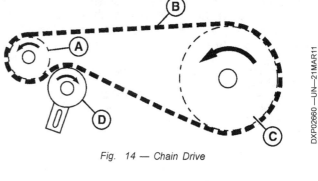

Fig. 14 — Chain Drive

A—Driving Sprocket   C—Driven Sprocket
B—Chain   D—Idler Sprocket (Adjustable)

Continued on next page

SPECIAL DRIVES

PLAIN OR DETACHABLE-LINK CHAIN

Once used more widely, plain chains (Fig. 15) are now used primarily for low-speed and light-load applications. Drag chains and flight elevators are common examples.

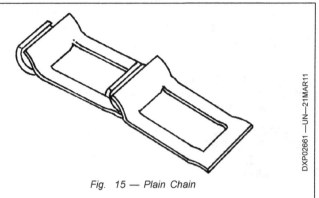

Fig. 15 — Plain Chain

ROLLER CHAINS

Roller chains (Fig. 16) are almost the universal choice for agricultural and construction equipment. They are much stronger than plain chains, and they last much longer. Roller bearings are built into every link.

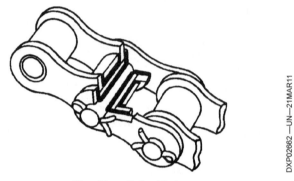

Fig. 16 — Roller Chain

Fig. 17 shows roller chain construction. Roller links alternate with pin links. Each roller link is an assembly of two rollers, two bushings, and two side bars. The rollers can rotate freely on the bushings. Pin links tie the roller links together.

If needed for strength, double-wide chains can be installed on double sprockets. For light loads, chains sometimes have extra-long side bars to engage only every second tooth on the sprockets.

A—Side Bar
B—Bushing
C—Roller
D—Side Bar
E—Roller Link
F—Pin
G—Pin Link
H—Links Assembled

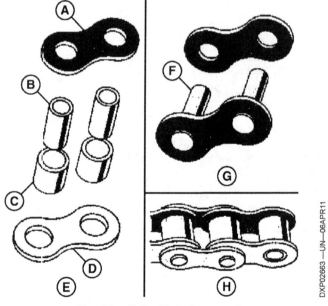

Fig. 17 — Roller Chain Components

Continued on next page

SPECIAL DRIVES

SILENT CHAINS

Silent chains (Fig. 18) are available for specialized applications such as an engine camshaft timing chain. Their sprockets are similar to gears. They can be run at higher speeds than roller chains.

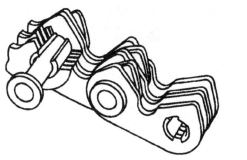

Fig. 18 — Silent Chain

Fig. 19 shows silent chain construction.

A—Pin
B—Main Link
C—Guide Link

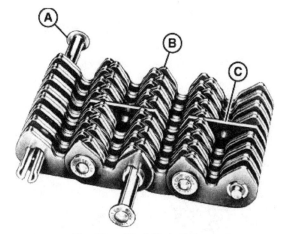

Fig. 19 — Silent Chain Components

Continued on next page

SPECIAL DRIVES

CHAIN DRIVE MAINTENANCE

Shafts need to be parallel (Fig. 20).

Sprocket alignment is critical (Fig. 21).

Fig. 20 — Aligning Shafts on a Chain Drive Using Machinist's Level

Fig. 21 — Aligning the Sprockets

Install a chain by positioning the ends on a sprocket and installing the connecting link (Fig. 22). After securing the side bar onto the connecting link, tap the pins back slightly. This prevents squeezing the roller links and allows lube to reach the bushings.

A—Connecting Link **B—Side Bar**

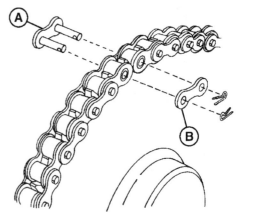

Fig. 22 — Installing the Chain

Continued on next page

SPECIAL DRIVES

Tension isn't as critical for chains as for belts. For most applications, just allow slight deflection on the slack side (Fig. 23). Vertical chains must not be allowed to hang away from the lower sprocket. A spring-loaded idler on the slack side will keep it snug. Chains should never be stretched too tight.

A—Adjust Tension at Idler

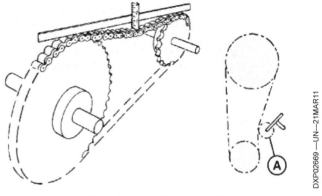

Fig. 23 — Checking Tension on Chains

After many hours of use, a roller chain develops slight looseness in each link. This allows the chain to "stretch," so the pitch no longer matches tooth spacing on the sprocket (Fig. 24). As a result, load is concentrated on a few teeth, accelerating wear of both chain and sprocket. Replace the chain before this happens.

Fig. 24 also shows a badly worn sprocket. Such wear disturbs the fit between chain and sprocket and can interfere with smooth entry and exit. The sprocket should be reversed, if possible, or replaced.

The ideal environment for a chain is inside a sealed housing, with the chain slightly submerged in oil at its lowest point. Such ideal conditions aren't often possible. Follow the manufacturer's instructions for cleaning, inspecting, and lubricating each chain.

For more extensive information, refer to Fundamentals of Service—Belts and Chains.

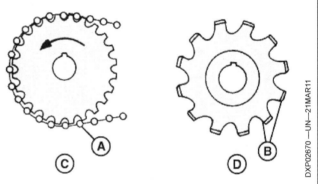

Fig. 24 — Problems with Chain Drives

A—Chain on Tips of Teeth
B—Hooked Teeth
C—Stretched Chain
D—Tooth Wear

RECIPROCATING DRIVES

Various machines require that rotary motion be converted to linear motion. From engine valves to cutterbars to baler plungers to piston pumps, we need ways to move parts back and forth.

CRANKWHEEL AND PITMAN LEVER DRIVE

One simple approach is to attach a pitman off-center on a wheel (Fig. 25). As the wheel rotates, the other end of the pitman moves a device back and forth.

A—Crankwheel
B—Pitman (Lever)
C—Knife

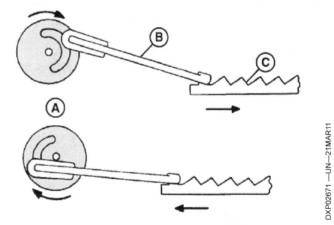

Fig. 25 — Crankwheel and Pitman Lever Driving a Mower Blade

Continued on next page

SPECIAL DRIVES

The basic pitman drive (Fig. 25) has certain disadvantages. It tends to vibrate due to imbalance, and the alternate pushing and pulling at an angle can cause side-to-side "slap" of the driven component. Fig. 26 shows how a swaybar eliminates the slap. Mounted on a fixed pivot, the swaybar can't be deflected.

A—Pitman
B—Crankwheel
C—Swaybar
D—Slight Arc

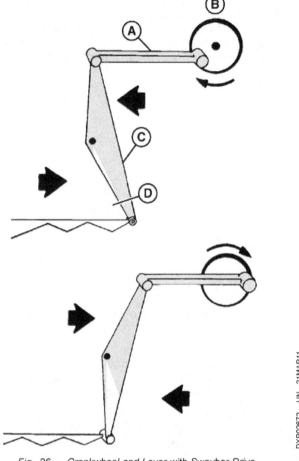

Fig. 26 — Crankwheel and Lever with Swaybar Drive

Fig. 27 shows twin crankwheels to reduce both vibration and slap. The crankwheels rotate in opposite directions. The two pitmans can be short, because their equal-and-opposite angles offset any side loading.

A—Twin Wheels
B—Short Pitman
C—Yoke
D—Blade

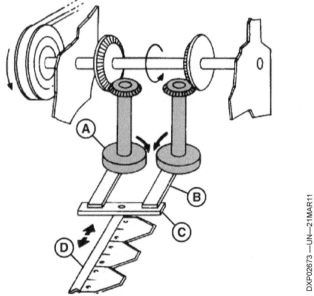

Fig. 27 — Twin Wheel Drive Operating Mowing Cutterbar

Continued on next page

SPECIAL DRIVES

CRANKARM AND LEVER DRIVE

A crankarm is equivalent to a crankwheel. Like a connecting rod journal on an engine crankshaft, it moves a pitman back and forth. Fig. 28 shows the plunger drive on a hay baler. Because there is heavy loading on the compression stroke and none on the return, a flywheel stabilizes the speed.

A—Pitman (Lever)
B—Work Element (Plungerhead)
C—Flywheel or Counterbalance
D—Drive Shaft
E—Bevel Gears
F—Crankarm

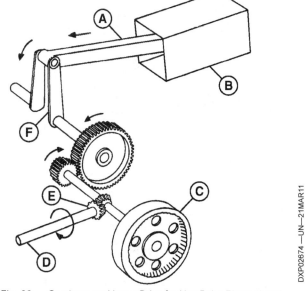

Fig. 28 — Crankarm and Lever Drive for Hay Baler Plungerhead

OFF-CENTER CRANKARM DRIVE

A "wobble shaft" (Fig. 29) also produces reciprocating motion. An angled crankarm rotates inside a sleeve. The ends of the sleeve move in a circular path. A hinged yoke absorbs the vertical motion, transferring the horizontal motion to a lever arm.

A—Drive Sheave
B—Wobble Shaft
C—Hinged Yoke
D—Lever Arm
E—Counterweight
F—Vertical Shaft
G—Crankarm

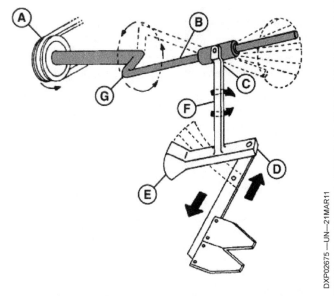

Fig. 29 — Off-Center Crankarm Drive (Wobble Shaft) for Mower Cutterbar

SPECIAL DRIVES

CAM DRIVES

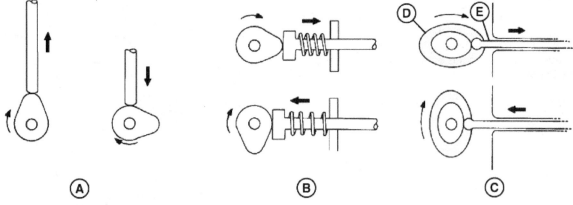

Fig. 30 — Cam Drives

A—Gravity Cam
B—Spring-Loaded Cam
C—Track and Follower Cam
D—Cam Track
E—Cam Follower

Camshafts are familiar mechanisms for converting rotary motion to reciprocating motion. Engine intake and exhaust valves are perhaps the most familiar example, but there are many others. Fig. 30 shows various configurations.

In most cases, a lobe on one side of the camshaft pushes a cam follower (left and center). If needed, the cam can be shaped to provide multiple or extended movements in each rotation.

In certain applications, the cam follower is captured in a cam "track" or groove (right). There is typically a roller on the cam follower to reduce friction. The cam track can be whatever pattern is needed. The transmission shifter cams shown in chapter 3 are specialized examples.

MAINTENANCE

Reciprocating drives are very machine-specific. Always refer to the correct manual for information on lubrication, inspection, alignment, etc.

SPECIAL DRIVES

OVERLOAD PROTECTION DEVICES

Many machines incorporate torque limiters of some sort. Unless the drive mechanism is robust enough to kill the engine without damage to the machine, it must be protected.

SLIP CLUTCHES

A slip clutch (Fig. 31) is a friction device that will slip if torque is too high. A friction disk is squeezed between two plates. Spring force must be adjusted so that the clutch will not slip in normal operation, but will slip before something breaks if the machine is overloaded.

Slippage can be just a momentary squeak, as might happen with harsh engagement of a heavy inertial load. Or it can be total stoppage, as when a foreign object blocks the machine.

Continuous slipping could quickly damage the slip clutch, so power should be disengaged immediately if a clutch begins to slip.

In Fig. 31, the input shaft is attached to the friction disk. The output shaft is attached to the outer plates. As on most slip clutches, maximum torque is adjustable. Always refer to the proper manual for slip clutch adjustment.

A—Adjusting Spring
B—Clutch Facing
C—Revolving Plate
D—Two-Piece Drive Shaft
E—Slip Clutch (Engaged)

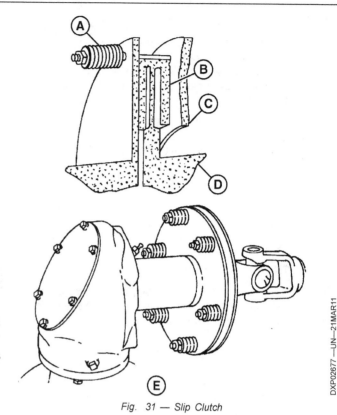

Fig. 31 — Slip Clutch

JUMP CLUTCHES

A jump clutch (Fig. 32) provides a variation on the same idea. Instead of a friction disk, it uses multiple wedges squeezed together by adjustable spring force. It slips if load is great enough to compress the spring by forcing the wedges apart.

A jump clutch makes a loud noise when it slips, but there is less risk of it being damaged.

A—Ratchet Clutch
B—Adjusting Spring
C—Disengaged
D—Engaged

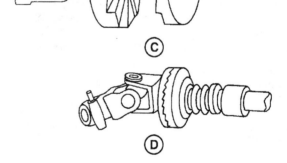

Fig. 32 — Jump Clutch

Continued on next page

SPECIAL DRIVES

SHEAR PINS

Shear pins (Fig. 33) are the simplest and least expensive overload protection devices. The biggest disadvantage is that they are not self-resetting. Like a blown fuse, a sheared pin must be replaced before the machine can be used.

It's important to use a shear pin of the correct size and hardness. Do not substitute a hardened steel bolt. Shear pins are much less expensive than the parts they are designed to protect.

A—Shear Pin
B—Gearbox
C—Drive Shaft

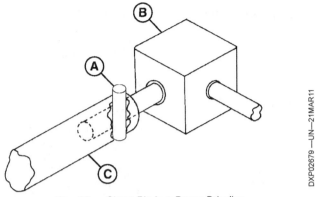

Fig. 33 — Shear Pin in a Power Driveline

TORSIONAL COUPLERS

The first component in many drive trains is a torsional coupler or damper (Fig. 34). Bolted to the flywheel, it provides a cushion between engine and transmission.

Fig. 34 — Torsional Coupler

The torsional coupler is a rubber ring (Fig. 35) bonded to an outer drum and an inner hub. The drum is attached to the flywheel, the hub to the drive shaft.

The torsional coupler protects the transmission from pulsations in the engine. Perhaps more important, it protects the engine from shock loading. As described in the previous chapter, U-joints in the PTO driveline can generate severe speed pulsations. A torsional coupler helps prevent feedback of such surges into the engine.

Almost the only failure mode for a torsional coupler is slippage of the hub or drum, usually the hub. Slippage quickly leads to total failure. It is not repairable.

A—Outer Drum
B—Drive Shaft
C—Bonded Rubber Cushion
D—Inner Hub

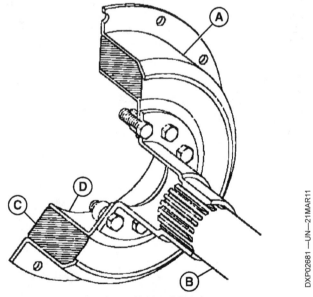

Fig. 35 — Torsional Coupler

SPECIAL DRIVES

SAFETY RULES

You must recognize potential hazards and take necessary action to avoid injury. Here are some safety rules for special drives.

A—Thread
B—In an Instant, Thread Wraps Around Shaft
C—Sleeve Is Immediately Pulled and Begins to Wrap

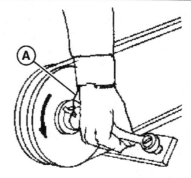

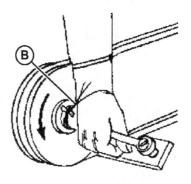

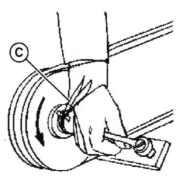

Fig. 36 — wrapping may begin with just a thread. In an instant, the victim is entangled with little chance to escape.

SPECIAL DRIVES

1. Do not wear loose clothing. Pinch points that are in belt drives, chain drives, and gear drives can catch clothing, pulling you into the machine, causing serious injury (Fig. 36).

2. Belts and chains may be under high tension; serious injury may occur if released unexpectedly.

3. Avoid freewheeling parts. Some parts can cause serious injury when coasting slowly to a stop. Watch for motion to stop (Fig. 37).

4. Keep guards and shields in place.

5. Never lubricate a chain or gear drive while the machine is running. Always shut off the engine and remove the key.

6. Replace all parts with exact duplicates. Refer to parts catalog before installing.

FAILURES OF BELTS

For details on failures of belts, see Fundamentals of Service—Belts and Chains. Or see FOS—Identification of Parts Failures.

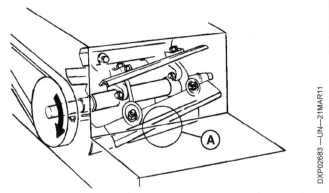

Fig. 37 — Even Parts That Are Moving Slowly Can Cause Serious Injury

A—Pinch/Shear Point

TROUBLESHOOTING OF BELT DRIVES

Following is a summary of trouble signs and possible causes of belt drive failures.

TROUBLESHOOTING OF BELT DRIVES	
Touble	Possible Cause
Belts turn over in sheaves	Misalignment of sheaves and shafts
	Worn sheave grooves
	Misalignment of flat idler sheave
	Excessive belt vibration
	Belt cord damage from wrong installation
Belt squeaks or screeches	Starting load too high, causing belt to slip because of improper tension
	Overload, causing belt to slip because of improper tension
Belt chirps	Movement of belt on flat idler or pulley (not harmful)
Belt stretched beyond take-up	Drive operated under too much tension
	Wrong belt or poor storage in damp area
Belt breaks prematurely	Foreign material in sheaves
	Shock or extreme overload
	Belt damaged during installation
Belt has short life	Worn sheaves
	Oil or grease on belt
	High temperatures
	Belt cover wear, caused by guard or guide interference
	Excessive belt slippage
	Poor storage of belts

SPECIAL DRIVES

TROUBLESHOOTING OF CHAIN DRIVES

MAINTENANCE

The problems of *gear wear* and *backlash* are also explained in chapter 1.

LUBRICATION

Gear lubricants range from simple mineral oils to complex formulas containing many ingredients.

For the type of oil to use, consult the manufacturer's operator's manual.

Common gear lubricants are:

Spur Gears — may use a good grade of petroleum oil — but not always.

Bevel Gears — may use an extreme-pressure (E.P.) lubricant.

Worm Gears — may use an SCL lubricant, which offers even greater lubricant qualities.

TROUBLESHOOTING OF CHAIN DRIVES

Trouble	Possible Cause
Excessive noise	Misalignment of sprockets
	Wrong tension
	Lack of lubrication
	Loose casings or bearings
	Chain or sprocket worn out
	Chain pitch too large
Wear on side bars, link plates, and sides of sprocket teeth	Misalignment
Chain climbing sprockets	Wrong chain or poor chain
	Chain worn out
	Lack of chain wrap on sprocket
	Too much chain slack
	Material buildup in sprocket tooth pockets
	Worn sprockets
Broken pins, bushings, or rollers	Chain runs too fast
	Heavy slack or sudden loads
	Material buildup in sprocket tooth pockets
	Lack of lubrication
	Chain or sprocket corrosion
	Wrong chain or worn sprockets
Chain clings to sprocket	Wrong chain or worn sprockets
	Heavy or tacky lubricant
	Excessive chain slack
	Material buildup in sprocket tooth pockets
Chain whips	Too much slack in chain
	High, pulsating loads
	Stiff chain joints
	Uneven wear on chain
Chain gets stiff	Lack of lubrication, resulting in wear
	Excessive overloads
	Corrosion
	Material buildup in chain joint
	Peening of side plate edges
	Misalignment
Broken sprocket teeth	Obstruction or foreign material
	Excessive shock loads
	Chain climbing sprocket teeth
Chain fasteners fail	Vibration
	Obstructions striking fasteners
	Fasteners improperly installed
Drive runs too hot	Chain running too fast
	Lack of lubrication
	Chain immersed too deep in oil bath
	Chain or shafts rubbing against obstruction

Continued on next page

SPECIAL DRIVES

TROUBLESHOOTING OF GEAR DRIVES

MAINTENANCE OF RECIPROCATING DRIVES

Lubrication is the key to good operation of all mechanical drives. Any moving part or parts making contact with another part should be well lubricated.

1. Bearings should be kept lubricated and checked for wear or damage periodically.
2. Cam tracks or cam surfaces require greasing for smooth operation and minimum wear.
3. Hinged areas need lubrication.
4. Enclosed drives should have oil level checked periodically.
5. Lever and work element must be aligned.
6. Above all, refer to the machine operator's manual for proper lubrication.

TROUBLESHOOTING OF GEAR DRIVES	
Trouble	Possible Cause
Fins around edge of teeth	Too much backlash
	Heavy, continuous loading
	Intermittent overloading
	Gears not hardened
Gear teeth appear burned from overheating	Too little backlash
	Overspeeding
	Overloading
	Lack of lubrication
Gear teeth are scored	Lack of proper lubrication
	Excessive speeds
	Overloading
	Too little backlash
Abrasions or fine tooth scratches	Dirt, grit, or metal particles in gears
Interference marks on tooth edges	Improper manufacturing
	Faulty assembly
Pitting of gear teeth	Excessive loading
Gouging near bottom of teeth	Too little backlash
Teeth broken	Too much backlash
	Shock impact or overload

SPECIAL DRIVES

TROUBLESHOOTING OF OVERLOAD MECHANISMS

MAINTENANCE OF OVERLOAD MECHANISMS

LUBRICATION

Lubrication is generally simple in an overload mechanism. Remember these points:

1. A light coating of oil can be used to keep clutch faces from sticking.

2. Too much oil on clutch faces reduces friction and causes slipping.

3. Spring tension for safety mechanisms may require light lubrication.

4. Shafts that are actually separated, but run inside a connecting shaft, should be lubricated to inhibit rust and corrosion that may prevent disengagement of the powerline.

TROUBLESHOOTING OF OVERLOAD MECHANISMS	
Trouble	Possible Cause
Slip clutch continually slips	Too little spring tension, Belleville washer worn out
	Oiled or glazed clutch facings
	Improper assembly
	Failure of the working element
Slip clutch will not slip	Facings have become stuck
	Too much spring tension
	Improper assembly
Jump clutch continually jumps	Too little spring tension
	Ratchet worn out
	Ratchet surface is slick from lubricant
	Dirt build-up not permitting complete contact between ratchet teeth
	Working element not functioning
	Drive overloaded
Jump clutch will not jump	Too much spring tension
	Spring bottomed out
	Ratchet frozen due to rust or corrosion
	Dirt preventing operation of spring release or ratchet
	Internal shaft frozen to external shaft
Shear pin continually shears	Wrong shear pin or bolt
	Working element not functioning
	Drive overload
Shear pin will not shear	Wrong shear pin or bolt
	Internal shaft frozen to external shaft

TEST YOURSELF

QUESTIONS

1. What are the three major kinds of special drives?

2. Which one is a friction drive?

3. How do the pulleys change belt speed in a variable-speed drive?

4. (True or False?) If one belt breaks in a multiple set, you need not replace all the belts.

5. Should the idler be on the tight or slack side of the chain?

6. Reciprocating drives change a _____ motion into a _____ motion.

(Answers in back of manual.)

DEFINITIONS OF TERMS AND SYMBOLS

DEFINITIONS

A

ABRASION — Wearing or rubbing away of a part.

ALIGNMENT — An adjustment to a line or to bring into a line.

ANTI-CLOCKWISE ROTATION — Rotating the opposite direction of the hands on a clock. The same as counterclockwise rotation.

ANTI-FRICTION BEARING — A bearing constructed with balls, rollers, or the like between the journal and the bearing surface to provide rolling instead of sliding friction.

ASAE — American Society of Agricultural Engineers.

ASME — American Society of Mechanical Engineers.

AXIAL — Parallel to the shaft or bearing bore.

AXLE — The shaft or shafts of a machine upon which wheels are mounted.

AUTOMATIC TRANSMISSION — A transmission in which gear or ratio changes are self-activated.

B

BACKLASH — The clearance or "play" between two parts, such as meshed gear teeth.

BALL BEARING — An anti-friction bearing consisting of hardened inner and outer races with hardened steel balls that roll between the two races.

BDC — Bottom dead center.

BEARING — The supporting part that reduces friction between a stationary and rotating part.

BENT-AXIS HYDROSTATIC — A hydrostatic pump and motor in which the rotating piston blocks are at an angle to the input and output shafts, as opposed to using swashplates. The angle can be fixed or variable.

BEVEL SPUR GEAR — Gear that has teeth with a straight centerline cut on a cone.

BONDED LINING — A method of cementing brake linings to shoes or bands, which eliminates the necessity of rivets.

BRAZE — To join two pieces of metal with the use of a comparatively high melting point material. An example is to join two pieces of steel by using brass or bronze as a solder.

BREAK-IN — The process of wearing-in to a desirable fit between the surfaces of two new or reconditioned parts.

BROACH — To finish the surface of metal by pushing or pulling a multiple-edge cutting tool over or through it.

BURNISH — To smooth or polish by the use of a sliding tool under pressure.

BUSHING — A removable liner for a bearing.

C

CARRIER — An object that bears, cradles, moves, or transports some other object or objects.

CASE-HARDEN — To harden the surface of steel.

CASTELLATE — Formed to resemble a castle battlement, as in a castellated or castle nut.

CENTER OF GRAVITY — The point at which a mass of matter balances. For example, the center of gravity of a wheel is the center of the wheel hub.

CENTRIFUGAL FORCE — A force that tends to move a body away from its center of rotation. Example: whirling a weight attached to a string.

CHAMFER — A bevel or taper at the edge of a hole or a gear tooth.

CHASE — To straighten up or repair damaged threads.

CHILLED IRON — Cast iron on which the surface has been hardened.

CLEARANCE — The space allowed between two parts, such as between a journal and a bearing.

CLUTCH — A device for connecting and disconnecting the engine from the transmission or for a similar purpose in other units.

COEFFICIENT OF FRICTION — The ratio of the force resisting motion between two surfaces in contact to the force holding the two surfaces in contact.

COMPOUND — A mixture of two or more ingredients or elements.

CONCENTRIC — Two or more circles, having a common center.

CONSTANT MESH TRANSMISSION — A transmission in which the gears are engaged at all times, and shifts are made by sliding collars, clutches, or other means to connect the gears to the output shaft.

CONTRACTION — A reduction in dimension; the opposite of expansion.

CORRODE — To eat away gradually as if by gnawing, especially by chemical action.

COUNTERBORE — To enlarge a hole to a given depth.

COUNTERCLOCKWISE ROTATION — Rotating in the opposite direction of the hands on a clock.

COUNTERSHAFT — An intermediate shaft that receives motion from a main shaft and transmits it to a working part. Sometimes called a lay shaft.

COUNTERSINK — To cut or form a depression to allow the head of a screw to go below the surface.

COUPLING — A connecting means for transferring movement from one part to another; may be mechanical, hydraulic, or electrical.

Continued on next page

DEFINITIONS OF TERMS AND SYMBOLS

D

DEAD AXLE — An axle that only supports the machine and does not transmit power.

DEFLECTION — Bending or movement away from normal due to loading.

DENSITY — Compactness; relative mass of matter in a given volume.

DIAGNOSIS — A systematic study of a machine or machine parts to determine the cause of improper performance or failure.

DIAL INDICATOR — A measuring instrument with the readings indicated on a dial rather than on a thimble as on a micrometer.

DIFFERENTIAL GEAR — The gear system that permits one drive wheel to turn faster than the other.

DIRECT DRIVE — Direct engagement between the engine and drive shaft where the engine crankshaft and the drive shaft turn at the same rpm.

DISTORTION — A warpage or change in form from the original shape.

DOUBLE REDUCTION AXLE — A drive axle construction in which two sets of reduction gears are used for extreme reduction of the gear ratio to reduce the speed.

DOWEL PIN — A pin inserted in matching holes in two parts to maintain those parts in fixed relation to one another.

DRIVELINE — The universal joints, drive shaft, and other parts connecting the transmission with the driving axles.

DROP FORGING — A piece of steel shaped between dies while hot.

DUAL REDUCTION AXLE — A drive axle construction with two sets of pinions and gears, either of which can be used.

E

ECCENTRIC — One circle within another circle wherein both circles do not have the same center. An example of this is a camshaft fuel pump lobe.

ENDPLAY — The amount of axial or end-to-end movement in a shaft due to clearance in the bearings.

F

FEELER GAUGE — A metal strip or blade finished accurately with regard to thickness used for measuring the clearance between two parts; such gauges ordinarily come in a set of different blades graduated in thickness by increments of 0.001 inch.

FILLET — A rounded filling between two parts joined at an angle.

FIT — The contact between two machined surfaces.

FLANGE — A projecting rim or collar on an object for keeping it in place.

FLUID DRIVE — A drive in which there is no mechanical connection between the input and output shafts, and power is transmitted by moving oil. (See chapter 6 on Torque Converters or chapter 5 on Hydrostatic Drives.)

FOOT-POUND (or ft.-lb.) — This is a measure of the amount of energy or work required to lift one pound a distance of one foot. Also used as a unit of torque equivalent to a one-pound force applied to a one-foot arm. (See TORQUE definition.)

FREEWHEELING CLUTCH — A mechanical device that will engage the driving member to impart motion to a driven member in one direction but not the other. Also known as an "overrunning clutch."

G

GEAR — A cylinder- or cone-shaped part having teeth on one surface that mate with and engage the teeth of another part that is not concentric with it.

GEAR RATIO — The ratio of the number of teeth on the larger gear to the number of teeth on the smaller gear.

GRIND — To finish or polish a surface by means of an abrasive wheel.

H

HAZARD — Dangerous object or situation that has the potential to cause injury.

HEAT TREATMENT — Heating, followed by fast cooling, to harden metal.

HEEL — The outside, larger half of the gear tooth.

HELICAL — Shaped like a coil spring or a screw thread.

HELICAL GEAR — Gears with the teeth cut at an angle to the axis of the gear.

HERRINGBONE GEAR — A pair of helical gears designed to operate together. The angle of the pair of gears forms a V.

HUB — The central part of a wheel or gear.

HYDRAULIC PRESSURE — Pressure exerted through the medium of a liquid.

HYPOID GEAR — A gear that is similar in appearance to a spiral bevel gear, but the teeth are cut so that the gears match in a position where the shaft centerlines do not meet.

I

ID — Inside diameter.

IMPELLER — The driving member or centrifugal pump in a torque converter.

Continued on next page

DEFINITIONS OF TERMS AND SYMBOLS

INFINITELY VARIABLE TRANSMISSION (IVT) — A transmission in which the output speed can be steplessly adjusted through the entire range. Also called continuously variable transmission (CVT).

INPUT SHAFT — The shaft carrying the driving gear by which the power is applied, as to the transmission.

J

JOURNAL — A smooth surface on a shaft or in a housing where a bearing is installed to support a rotating shaft.

K

KEY — A small block inserted between the shaft and hub to prevent circumferential movement.

KEYWAY — A groove or slot cut to permit the insertion of a key.

KNURL — To indent or roughen a finished surface.

L

LAPPING — The process of fitting one surface to another by rubbing them together with an abrasive material between the two surfaces.

LIMITED SLIP DIFFERENTIAL — Differential assembly designed to automatically balance driving power to the available traction at each wheel.

LINKAGE — Any series of rods, yokes, and levers, etc., used to transmit motion from one unit to another.

LIVE AXLE — A shaft that transmits power from the differential to the wheels and supports the weight of the machine.

LOST MOTION — Motion between a driving part and a driven part that does not move the driven part. Also see BACKLASH.

LOW SPEED — The gearing that produces the highest torque and lowest speed of the wheels.

M

MISALIGNMENT — Condition when bearings, shafts, sprockets, pulleys, etc., are not in a straight line.

MODULATION — Smooth engagement of a clutch to avoid shock loading or harsh starts. Term also applies to adjustment of speed, pressure, etc.

MULTIPLE DISK — A clutch with a number of driving and driven disks as compared to a single plate clutch.

N

NEEDLE BEARING — An anti-friction bearing using a great number of long, small-diameter rollers. Also known as a quill bearing.

O

OD — Outside diameter.

OSCILLATE — To swing back and forth like a pendulum.

OUTPUT SHAFT — The shaft or gear that delivers the power from a device, such as a transmission.

OVERDRIVE — Any arrangement of gearing that produces more revolutions of the driven shaft than of the driving shaft.

OVERRUN COUPLING — A freewheeling device to permit rotation in one direction but not in the other.

P

PEEN — To stretch or clinch over by pounding with the rounded end of a hammer.

PINION — The smaller of two meshing gears.

PINION CARRIER — The mounting or bracket that retains the bearing supporting a pinion shaft.

PLANETARY GEAR SET — A system of gearing that is modeled after the solar system. A sun pinion is surrounded by an internal ring gear, and planet gears are in mesh between the ring gear and sun pinion, around which all revolve.

PLANET CARRIER — In a planetary gear system, the carrier or bracket in a planetary system that contains the shafts upon which the pinions or planet gears turn.

PLANET GEARS — The gears in a planetary gear set that connect the sun gear to the ring gear.

POWER SHIFT TRANSMISSION — A transmission in which gear changes are selected manually, but are power actuated.

PRELOAD — A load within the bearing, either purposely built in or resulting from adjustment.

PRESS FIT — Mounting with interference, e.g., bore of bearing is smaller than OD of shaft, or OD of bearing is larger than bore of housing, or both.

PULSE WIDTH MODULATION (PWM) — A means of controlling electrohydraulic valves by switching the voltage on and off many times per second. Voltage and frequency remain the same, but the percentage of time the voltage is on can be adjusted to regulate pressure.

PUMP — A device that produces motion in a liquid. In a torque converter, the driving member. (Also see IMPELLER.)

R

RACE — A channel in the inner or outer ring of an anti-friction bearing in which the balls or rollers roll.

RADIAL — Perpendicular to the shaft or bearing bore.

RADIAL CLEARANCE (Radial displacement) — Clearance within the bearing and between balls and races perpendicular to the shaft.

RADIAL LOAD — A force perpendicular to the axis of rotation.

Continued on next page

DEFINITIONS OF TERMS AND SYMBOLS

RATIO — The relation or proportion that one number bears to another.

REAM — To finish a hole accurately with a rotating fluted tool.

RECIPROCATING — A back-and-forth movement, such as the action of a piston in a cylinder.

RING GEAR — A gear that surrounds or rings the sun and planet gears in a planetary system. Also the name given to the spiral bevel gear attached to a differential.

RIVET — A headed pin used for uniting two or more pieces by passing the shank through a hole in each piece, and securing it by forming a head on the opposite end.

ROLLER BEARING — An inner and outer race upon which hardened steel rollers operate.

RPM — Revolutions per minute.

S

SAE — Society of Automotive Engineers.

SCORE — A scratch, ridge, or groove marring a finished surface.

SEAT — A surface, usually machined, upon which another part rests or seats; for example, the surface upon which a valve face rests.

SEPARATORS — A component in an anti-friction bearing that keeps the rolling components apart.

SHIM — Thin sheets used as spacers between two parts, such as the two halves of a journal bearing.

SHRINK-FIT — Where the shaft or part is slightly larger than the hole in which it is to be inserted. The outer part is heated above its normal operating temperature, or the inner part chilled below its normal operating temperature, or both, and they are assembled in this condition. Upon cooling, an exceptionally tight fit is obtained.

SLIDING-FIT — Where sufficient clearance has been allowed between the shaft and journal to allow free-running without overheating.

SLIDING GEAR TRANSMISSION — A transmission in which gears are moved on their shafts to change gear ratios.

SLIP-IN BEARING — A liner made to extremely accurate measurements, which can be used for replacement purposes without additional fitting.

SPIRAL BEVEL GEAR — A ring gear and pinion wherein the mating teeth are curved and placed at an angle with the pinion shaft.

SPIRAL GEAR — A gear with teeth cut according to a mathematical curve on a cone. Spiral bevel gears that are not parallel have centerlines that intersect.

SPLINE — Splines are multiple keys in the general form of internal and external gear teeth, used to prevent relative rotation of cylindrically fitted parts.

SPUR GEAR — Gears cut on a cylinder, with the teeth are straight and parallel to the axis.

SQ. FT. — Square feet

SQ. IN. — Square inch

STALL CONDITION — A condition in a torque converter when the driving element (pump) is turning and the driven element (turbine) is stopped. Stall produces maximum vortex flow. (See chapter 6 on Torque Converters.)

STATOR — In a torque converter, the third member (in addition to turbine and pump) that changes direction of fluid under certain operating conditions.

STRESS — The force to which a material, mechanism, or component is subjected.

SUN GEAR — The central gear in a planetary gear system around which the rest of the gears rotate.

SYNCHROMESH TRANSMISSION — See CONSTANT MESH TRANSMISSION.

SYNCHRONIZE — To cause two events to occur at the same time. For example, to bring two gears to the same speed before they are meshed.

T

TAP — To cut threads in a hole with a tapered, fluted, threaded tool.

TEMPER — To change the physical characteristics of a metal by applying heat.

TENSION — Effort that elongates or "stretches" a material.

THRUST LOAD — A load that pushes or reacts through the bearing in a direction parallel to the shaft.

TOLERANCE — A permissible variation between the two extremes of a specification or dimension.

TORQUE — A twisting force, usually measured in ft.-lb. (N·m). (See FOOT-POUND definition.)

TORQUE CONVERTER — A turbine device utilizing a rotary pump, stators, and one or more driven circular turbines or vanes whereby power is transmitted from a driving to a driven member by hydraulic action. It provides varying drive ratios; with a speed reduction, it increases torque.

TORQUE WRENCH — A special wrench with a built-in indicator to measure the applied force.

TORUS SECTION — The confines of a flow circuit in a radial plane in a torque converter or fluid coupler.

TRANSAXLE — Type of construction in which the transmission, differential, and axles are combined in one unit.

Continued on next page

DEFINITIONS OF TERMS AND SYMBOLS

TRANSMISSION — An assembly of gears or other elements that gives variations in speed or direction between the input and output shafts.

TROUBLESHOOTING — A process of diagnosing the source of the trouble or troubles through observation and testing.

TUNE-UP — A process of accurate and careful adjustments to restore the best performance.

TURBINE — A rotary device for obtaining mechanical power from a high-velocity flow of gases or liquids.

TURBULENCE — A disturbed or irregular motion of fluids or gases.

V

VANES — Any plates, blades, or the like attached to an axis and moved by air or a liquid.

VORTEX — A whirling movement or mass of liquid or air.

W

WORM GEAR — A gear with teeth that resemble a thread on a bolt. It is meshed with a gear that has teeth similar to a helical tooth except that it is dished to allow more contact.

MEASUREMENT CONVERSION CHART

MEASUREMENT CONVERSION CHART	
METRIC TO ENGLISH	ENGLISH TO METRIC
LENGTH	
1 millimeter = 0.03937 inches (in.)	1 inch = 25.4 millimeters (mm)
1 meter = 3.281 feet (ft.)	1 foot = 0.30048 meters (m)
1 kilometer = 0.621 miles (mi.)	1 mile = 1.608 kilometers (km)
AREA	
1 meter^2 = 10.76 square feet (sq. ft.)	1 square foot = 0.0929 meter^2 (m^2)
1 hectare = 2.471 acres (acre)	1 acre = 0.4047 hectare (ha)
1 hectare = 10,000 m^2	1 acre = 43,560 sq. ft.
MASS (WEIGHT)	
1 kilogram = 2.205 pounds (lb.)	1 pound = 0.4535 kilograms (kg)
1 tonne (1,000 kg) = 1.102 ton (tn)	1 ton (2,000 lb.) = 0.9071 tonnes (t)
VOLUME	
1 meter^3 = 35.31 cubic feet (cu. ft.)	1 cubic foot = 0.02832 meter^3 (m^3)
1 meter^3 = 1.308 cubic yards (cu. yd.)	1 cubic yard = 0.7646 meter^3 (m^3)
1 meter^3 = 28.38 bushel (bu.)	1 bushel = 0.03524 meter^3 (m^3)
1 liter = 0.02838 bushel (bu.)	1 bushel = 35.24 liters (L)
1 liter = 1.057 quart (qt.)	1 quart = 0.9464 liter (L)
1 liter = 0.2642 gallon (gal.)	1 gallon = 3.785 liters (L)
PRESSURE	
1 kilopascal = 0.145 pounds per square inch (psi)	1 psi = 6.895 kilopascals (kPa)
1 bar = 101.325 kilopascals	1 psi = 0.06895 bars (bar)
STRESS	
1 megapascal = 145 pounds per square inch (psi)	1 psi = 0.006895 megapascal (MPa)
1 megapascal = 1 newton/millimeter^2 (N/mm^2)	1 psi = 0.006895 newton/millimeter^2 (N/mm^2)
POWER	
1 kilowatt = 1.341 horsepower (hp.)	1 horsepower (550 ft.-lb./s) = 0.7457 killowatt (kW)
1 watt = 1 N/s	1 horsepower = 746 watts (W)
ENERGY (WORK)	
1 joule + 0.0009478 British thermal units (Btu)	1 British thermal unit = 1055 joules (J)
FORCE	
1 newton = 0.2248 pounds force (lb.-force)	1 pound = 4.448 newtons (N)
TORQUE OR BENDING MOMENT	
1 newton-meter = 0.7376 lb.-ft.	1 lb.-ft. = 1.356 newton-meters (N·m)
TEMPERATURE	
°C x 1.8 + 32 = 1°F	(°F − 32) / 1.8 = 1°C

ANSWERS TO TEST YOURSELF QUESTIONS

ANSWERS TO CHAPTER QUESTIONS

ANSWERS TO CHAPTER 1 QUESTIONS

1. a — 2; b — 3; c — 1.
2. **Gear B has traveled 1/2 revolution.**
3. **False.** To reduce speeds, a small gear drives a larger one.
4. **Reducing gear speeds increases the twisting force or torque.**
5. **True.**
6. **Friction, gears, and fluids.**
7. **Friction —** wheels, belts, clutches, etc. **Gears —** transmissions chains, etc. **Fluids —** water wheel, torque converter, fluid drive, etc.
8. Any one of these: **Operate more quietly, wear less rapidly, have greater tooth strength.**
9. **Reduce friction and support a shaft.**
10. **Preloaded.**
11. **End play.**
12. **Backlash.**

ANSWERS TO CHAPTER 2 QUESTIONS

1. Any four of these: **Disk and plate, band, overrunning, cone, expanding shoe, fluid, magnetic.**
2. **Wet and dry.**
3. Any two of these: **Long wearing, heat resistant, allow little slippage.**
4. **The flexible disk has springs around its hub.**
5. First blank — **engage.** Second blank — **freewheel.**
6. **False.** Over-center linkage locks both while engaged and disengaged.
7. **False.**
8. **C. Both of the above.**

ANSWERS TO CHAPTER 3 QUESTIONS

1. **Sliding gear, collar shift, and synchromesh.**
2. **True.**
3. **The collar shift transmission usually runs quieter than a sliding gear type.** (This is because it can use helical — instead of spur-tooth gears.)
4. **Synchromesh.**
5. **The countershaft serves to transmit and vary the power flow between the input and output shafts.**
6. **Block engagement of gear if speeds do not match; accelerate or decelerate one component to achieve equal speeds; allow engagement once speeds match.**

ANSWERS TO CHAPTER 4 QUESTIONS

1. First blanks — **hydraulic clutches.** Second blanks — **gear train.**
2. **Countershaft and planetary.**
3. **Underdrive means that the output shaft turns slower than the input, while overdrive means that the output turns faster than the input.**
4. **Sun gear, planetary pinions (and carrier), and ring gear.**
5. **True.**
6. **The ratio is 1 to 1 (a direct drive).**
7. **Lubricates (and cools); engages the clutches and brakes.**

ANSWERS TO CHAPTER 5 QUESTIONS

1. First blank — **low.** Second blank — **high.**
2. **A torque converter uses fluids at high speeds but relatively low pressures.**
3. **Pump and motor.**
4. **Variable.**
5. **The charge pump makes up for oil lost from the closed circuit between the pump and motor.**
6. **Steerable motor and non-steerable.**
7. **Overlapping of power and return strokes.**
8. **To pressurize the outer case and hold the pistons away from the cam lobes to disengage the motor.**

ANSWERS TO CHAPTER 6 QUESTIONS

1. First two blanks — **fluid coupling.** Second two blanks — **torque converter.**
2. **The stator.**
3. First blank — **vortex.** Second blank — **rotary.**
4. **False.**
5. **False.**
6. a — 3; b — 1; c — 2.

ANSWERS TO CHAPTER 7 QUESTIONS

1. **a.** There are other advantages, but the primary reason or reducing engine speed is to reduce fuel consumption.
2. **a.** Engine speed should increase to 1950 for maximum power before travel speed is reduced.

Continued on next page

ANSWERS TO TEST YOURSELF QUESTIONS

3. **c.** If the tractor is unable to pull the load at the desired speed, it should maintain 1950 rpm for maximum power and adjust transmission ratio pull the load as fast as possible.

4. **b.** IVT is much less complex than PST.

5. **b.** The hydrostatic unit is one of two inputs to the planetary.

6. **c.** PTO work should be done at rated speed. Certain other jobs should also be done at constant engine speed.

7. **b.** The shift occurs when sun gear and ring gear are turning at the same speed, so the entire planetary assembly rotates as one piece.

8. **c.** None of the above. Everything in IVT is controlled electronically.

9. **a.** By noting the sequence in which the two channels turn on and off, the computer senses direction of rotation.

10. **c.** Because the bent-axis hydrostatic works efficiently over a wide speed range, only two gears are needed.

ANSWERS TO CHAPTER 8 QUESTIONS

1. **Transmits power "around the corner" and allows each wheel to rotate at a different speed.**

2. **They lock out differential action to prevent a loss of power when one wheel slips.**

3. **Mechanical, hydraulic, and automatic (no-spin).**

4. **To prevent the danger of turning the machine with the differential lock engaged.**

5. **The automatic (no-spin) type.**

6. **False.** The faster wheel turns faster than ring gear speed, but the slower wheel turns the same as the ring gear.

ANSWERS TO CHAPTER 9 QUESTIONS

1. **Straight axle, pinion, planetary, and chain.**

2. **Straight axle.**

3. **A semi-floating axle transmits the torque as well as supporting part of the machine's weight. A full-floating axle only transmits torque.**

4. **False.**

5. Either of these: **More compact, more durable.**

6. The main advantage: **Allows for high clearance under the axle.** The main disadvantage: **Tends to loosen chain tension.**

7. **False.**

8. **No.**

ANSWERS TO CHAPTER 10 QUESTIONS

1. **a — 1; b — 3; c — 2.**

2. **They have a different number of splines (540 has 6 splines, while 1,000 has 21 splines).**

3. **True.**

ANSWERS TO CHAPTER 11 QUESTIONS

1. **Belt, chain, and gear.**

2. **Belt.**

3. **The two halves of the pulley move closer together or farther apart to move the belt in or out.**

4. **False.**

5. **On the slack side.**

6. First blank — **rotary.** Second blank — **linear or straight.**

Index

A

Adjusting the Gear Train
   Backlash in Gears 1-27
   Endplay in Gears and Shafts 1-27
   Preloading of Gear Trains 1-27
Adjustments
   Manual Transmissions 3-22
Air Clutches
   Other Types of Clutches 2-29
ASAE Standards (PTO) 10-4

B

Backlash in Gears
   Adjusting the Gear Train 1-27
   Gears 1-11
Band Clutches
   Other Types of Clutches 2-29
Bearings
   Ball Bearings 1-23
   Bearing Loads 1-23
   Needle Bearings 1-23
   Roller Bearings 1-23
Brake
   PTO 10-11

C

Cam Shifters 3-20
Capacity
   Design Considerations (Clutch) 2-4
Centrifugal Clutches
   Other Types of Clutches 2-29
Clutch Linkage
   Design Considerations (Clutch) 2-4
Clutches in Other Locations 2-22
Collar Shift Shifter 3-2
Cone Clutches
   Other Types of Clutches 2-29
Constant-Velocity Joints (PTO Controls) 10-12
Controls
   Contant-Velocity Joints 10-12
   Front PTO 10-12
   Lock-Out Devices 10-12
   Modulation 10-12
   Proper Timing of U-Joints 10-12
   PTO Drives 10-12
   Shielding 10-12
   Universal Joints 10-12

D

Design Considerations (Clutch)
   Capacity 2-4
   Clutch Linkage 2-4
   Disk 2-4
   Dual Clutch Assemblies 2-4
   Flywheel 2-4
   Pressure Plate Assembly 2-4
   Transmission Input Shaft 2-4
Design Variations (PTO) 10-5
Disk
   Design Considerations (Clutch) 2-4
Disk Clutch 2-2
Dual Clutch Assemblies
   Design Considerations (Clutch) 2-4

E

Endplay in Gears and Shafts 1-27
Expanding Shoe Clutches
   Other Types of Clutches 2-29

F

Flywheel
   Design Considerations (Clutch) 2-4
Front PTO (Controls) 10-12

G

Gear Ratios 1-11
Gear Wear 1-11
Gears
   Backlash in Gears 1-11
   Gear Ratios 1-11
   Gear Wear 1-11
   Planetary Gears 1-11
   Summary 1-11
   Types of Gears 1-11
General Maintenance
   Manual Transmissions 3-22

H

How a Power Train Works 1-2
How Power Is Transmitted
   Fluid Drives 1-8
   Friction Drives 1-8
   Gear Drives 1-8
Hydraulic Clutches
   Other Types of Clutches 2-29

I

Introduction
   Clutches 2-1
   Manual Transmission 3-1
   Power Take-Offs 10-1
   Power Trains — How They Work 1-1

Continued on next page

Index

L

	Page
Lock-Out Devices (PTO Controls)	10-12

M

Magnetic Clutches	
Other Types of Clutches	2-29
Modulation (PTO Controls)	10-12

O

Operation (PTO)	10-3
Other Types of Clutches	
Band Clutches	2-29
Centrifugal Clutches	2-29
Cone Clutches	2-29
Expanding Shoe Clutches	2-29
Hydraulic Clutches	2-29
Magnetic Clutches	2-29
Overrunning Clutches	2-29
Pneumatic (Air) Clutches	2-29
Slip Clutches	2-29
Overrunning Clutches	
Other Types of Clutches	2-29

P

Park Lock	3-22
Planetary Gears	1-11
Plate Clutch	2-2
Pneumatic Clutches	
Other Types of Clutches	2-29
Power Train Safety	1-30
Preloading of Gear Trains	1-27
Pressure Plate Assembly	
Design Considerations (Clutch)	2-4
Proper Timing of U-Joints (PTO Controls)	10-12
PTO Brake	10-11
PTO Drives (Controls)	10-12

S

Safety	
Power Train	1-30
Safety Rules for PTO	10-17
Service Notes	
Clutches in Other Locations	2-22
Shielding (PTO Controls)	10-12
Shift Controls	3-18
Sliding Gear Shifter	3-2
Slip Clutches	
Other Types of Clutches	2-29
Summary	
Power Trains — How They Work	1-33
Synchronizer Shifter	3-2

T

Test Yourself	
Clutches	2-37
Manual Transmissions	3-24
Power Trains — How They Work	1-35
PTO	10-17
Transmission Input Shaft	
Design Considerations (Clutch)	2-4
Troubleshooting	
Chattering (Clutch)	2-36
Clutches	2-36
Dragging (Clutch)	2-36
Failure (Clutch)	2-36
Grabbing (Clutch)	2-36
Manual Transmissions	3-23
PTOs	10-17
Rattles (Clutch)	2-36
Slipping (Clutch)	2-36
Squeaks (Clutch)	2-36
Summary (Clutches)	2-36
Vibrations (Clutch)	2-36
Types of Clutches	
Disk and Plate Clutches	2-2
Types of Gears	1-11
Types of PTO	10-2
Types of Shifters	
Collar Shift	3-2
Sliding Gear	3-2
Synchronizer	3-2

U

Universal Joints (PTO Controls)	10-12

POWER TRAINS

Power Trains is the definitive "how-to" book of offroad power train systems—from showing you how to diagnose problems and text components to explaining how to repair the system. And when we say "show you," we mean just that! Our book is filled with illustrations to clearly demonstrate what must be done... photographs, drawings, pictorial diagrams, troubleshooting charts, and diagnostic charts.

Instructions are written in simple language so that they can be easily understood. This book can be used by anyone, from a novice to an experienced mechanic.

By starting with the basics, the book builds your knowledge step by step. Chapter 1 covers how power trains work. Chapters 2–11 go into detail about the different kinds of power trains and their working parts. Each chapter clearly discusses how to adjust and maintain its power train system as well as how to diagnose and test problem areas.

JOHN DEERE
Deere & Company

FOS4008NC
ISBN 0-86691-377-7